Introduction

"From our heart to you, this book is for everyone to learn"

We are the Kupa Piti Kungka Tjuta - Senior Aboriginal Women of Coober Pedy, South Australia.

Us Kungkas, we are always talking strong. Never stop talking. Always thinking about the future generations, black and white.

Now we've stopped the radioactive waste dump we are having a good break. Still looking after the country and all the kids, that we never stop. Still teaching them our story. Until we finish we will tell our story.

This book is for everyone to learn. We give all our ideas about looking after the country. All the words from our heart, talking true way about everything we did fighting the poison. All the stories with all the pictures of us talking straight out, travelling everywhere.

Your turn now to look after the country properly. Don't forget about that poison. Be strong like us. Don't be scared of the Government. We weren't scared and we're elderly ladies.

From our heart to you, this book is for everyone to learn.

Ivy Makinti Stewart, Eileen Kampakuta Brown, Eileen Unkari Crombie, Emily Munyungka Austin, Tjunmutja Myra Watson.

Editorial Committee

Kupa Piti Kungka Tjuta

Coober Pedy 2005

Editors note

While packing up the Irati Wanti office we realised the campaign archives couldn't end up locked away, collecting dust in Coober Pedy. A book with 'all the words' and 'all the photos' seemed the right way to honour the Kungkas' efforts to teach everyone how to look after the country and culture.

Alapalatja Press is a self-publishing project by the Kungkas (Kupa Piti Kungka Tjuta) and the Browns (two greenie sisters) in Coober Pedy. The Kungkas often say "*ala palatja*—there you go!" after explaining something and are happy we've understood. Making a 'proper' book for the first time reflects the way the Irati Wanti campaign has worked; to listen and learn as you go.

This book is a collection of stories that have been documented over the years. It is not meant to be an in-depth account of the campaign and between each story are many more.

We would like to thank

All the Elders and staff from Umoona Aboriginal Aged Care, especially the Board of Management and Norman Riessen (Chairperson), Leigh Cleghorn (Manager) and Sonia Mazzone (Kungka Major).

Eve Vincent and Suzanne Woolford, our editorial assistants, for guiding us through the mysteries of proper English and offering vital outside perspectives.

Esther Singer, our design assistant, for her valuable design contributions and technical know how.

Samantha Sowerwine, Camilla Pandolfini and Sarah Heyward for proofreading.

Sister Michele Madigan for providing insight into the earlier years.

Everyone who gave us photos, memories, information and ideas.

And to our family and friends for keeping us on track. Always.

Nina Brown and Clare Brown
Coober Pedy 2005

Contents

Meet the Kupa Piti Kungka Tjuta   6

All the women talking up strong   7

Seven Sisters Dreaming   8

1998   9

1999   19

2000   27

2001   41

2002   49

2003   57

2004   101

July–August 2004   109

Image credits   118

Desarae Brown, Tiffany Brown, Georgia Brown. Dot design on banner by Desarae.

Language note

Yankunytjatjara is one of many dialects of the Western Desert language. Please note that Yankunytjatjara words and English translations throughout the book have been provided by the Kupa Piti Kungka Tjuta and are not necessarily the translations found in the *Pitjantjatjara/Yankunytjatjara To English Dictionary*, IAD Press.

All Yankunytjatjara words are translated except for:
A̲nangu—Aboriginal people
Inma—traditional song/dance

Meet the Kupa Piti Kungka Tjuta
Senior Aboriginal Women's Council of Coober Pedy

In the late 1980s Yankunytjatjara/Antikarinya Elders Ivy Makinti Stewart, Eileen Unkari Crombie and Emily Munyungka Austin were living in the outback opal mining town of Coober Pedy, in South Australia's far-north. The three *kungkas*—women were becoming very worried about the loss of the traditional culture, especially for the younger A*n*angu women living in town.

To share important women's cultural knowledge, Ivy Stewart, Eileen Crombie and Emily Austin founded the Kupa Piti Kungka Tjuta - Senior Aboriginal Women's Council of Coober Pedy. Alongside many local Aboriginal women, the Kungkas worked tirelessly throughout the 1990s with little or no resources to fulfil important cultural obligations. The Kungkas recruited Sister Michele Madigan as their 'honorary paperworker', and after much organisation they were officially incorporated in 1995.

"We take our responsibilities very seriously towards..."

- the land, the country, some of the special places, we know them
- the *Tjukur*—the important stories of the land
- the songs that prove how the land is
- the *Inma*—song and dance of the culture, all part of the land as well
- the bush tucker that we know and do our best to teach the grandchildren, and even tourists when we have the chance
- preserving the traditional crafts; the *wira*—wooden bowl, *wana*—digging stick, *punu*—music sticks, and even *kali*—boomerang, that our grandmothers have passed down to us through the generations
- the language
- the family, that members have respect for one another

"All this is law"

"We know what we are doing and what we are trying to do is very important. We don't want the culture to die. We want it to give strength to the land and also strength to ourselves, to our children and grandchildren who will have something more than video games, drinking and drugs and walking the street... And we know that our Aboriginal culture is very important, not just for A*n*angu but for our beautiful country, Australia."

Kungka tjutangku kunpu wangkanyi—All the women talking up strong

Over the years the Kupa Piti Kungka Tjuta Aboriginal Corporation included many women. Pictured below are the members that became active in the Irati Wanti campaign.

Ivy Makinti Stewart
Born: Iwantja

Eileen Unkari Crombie
Born: Sailors Well, near Marla Bore

Emily Munyungka Austin
Born: Apara

Eileen Kampakuta Brown
Born: Iltur

Eileen Wani Wingfield
Born: Ingomar

Angelina Wonga
Born: Everard Park

Martha Uganbari Edwards
Born: Tarcoola

Tjunmutja Myra Watson
Born: Between Coffin Hill and Maralinga

Betty Nyangala Muffler
Born: Wingelinna

Peggy Tjingila Cullinan
Born: Wamitjara
(Dec. 2000)

Lucy Kampakuta Wilton
Born: Mulytjantu
(Dec. 2001)

Pingkai Upitja
Born: Iltur
(Dec. 1998)

There were many other women who supported the Kungkas over the years including, Angelina Scobie, Nelly Mindum, Dianne Edwards, Eva Williams, Lois Brown, Rebecca Bear-Wingfield, Renee Wintinna, Charmaine Wells, Audrey Wintinna.

Tjukurpa Kungka Tjuta—Seven Sisters Dreaming

"They are in the stars reminding us"

It is the Dreaming from long, long ago.

The Seven Sisters travelled all over the country. Western Australia, New South Wales, South Australia, everywhere. One *wati*—man was always following behind while the sisters were travelling along. He was always trying to get one of the Seven Sisters, but the women are tricky, always looking out for the *wati*, travel all over, look in every tree, always on the lookout.

Women are always on the lookout. Lookout for the kids, for the country, for anything that might be wrong. Women come to be strong now because of the Seven Sisters story, the strongest to learn and teach. We are the ones willing to teach the young ones.

The *wati* caught one sister, the youngest with the boomerang, and the other sisters flew up into the sky after her. You can look into the sky and see the Seven Sisters still today. They are there in the stars, reminding us to be strong. We are always looking to the future, and this story is remembering.

We are strong today because of the *Tjukur*—Dreaming. We learnt it from the grandmothers, always following their footsteps. We sing their *Inma*, just as they sang in the beginning.

The Seven Sisters put everything in the *manta*—earth in the beginning, the *kungkas*—women's sacred *manta*.

Seven Sisters sitting around a waterhole
Image from an original painting by Emily Austin.

1998

A radioactive waste dump for SA?

Out of sight, out of mind

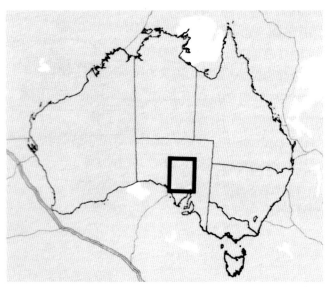

Spanning most of central-north South Australia, the Billa Kalina region includes the Woomera Prohibited Area, the townships of Roxby Downs, Andamooka and Woomera, and an extensive network of salt lakes. The region is very close to the Great Artesian Basin and Coober Pedy.

February 1998. The Federal Government announces their plan to build a national radioactive waste dump in South Australia. After shortlisting eight regions across Australia they identify Billa Kalina, a 67,000 square km region in SA's arid north, as the preferred area to build the dump.

Eighteen sites across the Billa Kalina region are selected for test drilling.

The Federal Government's Department of Industry, Science and Resources proposes the co-location of two separate facilities within the Billa Kalina region.

One: a shallow burial dump to bury low-level waste; radioactive for up to 300 years.

Two: an above ground facility to store intermediate long-lived waste; a serious radiation hazard for up to 10,000 years, requiring isolation for up to 250,000 years.

Lucas Heights - the driving force

A majority of the radioactive waste destined for the dump is produced at the Lucas Heights nuclear reactor in suburban Sydney. It has been operating since 1955; in 1997 the Federal Government announced plans to spend $300 million to replace the ageing reactor. The expansion would generate waste for another 40 years.

To reduce public opposition and gain a government licence for the new reactor, the Australian Nuclear Science and Technology Organisation (ANSTO) must present a convincing waste management plan. They ignore their own scientific experts who warn against shallow burial, pushing ahead with an 'out of sight, out of mind' approach.

Stockpiles of radioactive waste stored at Lucas Heights.

"We said NO straight away"

March

At one of their weekly meetings the Kungkas learned of the dump plan via a letter written to the *Coober Pedy Regional Times* from local Phil Gee. Eileen Crombie remembers the meeting, "We went wild. Straight out we said *wanti*—leave it. We were all saying it must be deadly poison if they want to bring it all the way over here, and we were frightened properly."

She recalls many of the women talking immediately about their experiences of the atomic tests carried out at Emu Junction and Maralinga in the 1950s and 60s. "We've had enough poison, enough sickness, from the bomb. We knew enough about the *irati*—poison from when we were young girls. We knew we had to fight it."

The Kungkas discussed "which way we were going to stop them [the Government] and who was going to help us". They suggested many people in town and surrounding areas they would contact hoping they would join in the fight against the dump.

"We kept saying no all the way."

"Listen to us. The desert lands are not as dry as you think."

Kungkas not sold on dump

April

Dr Caroline Perkins, Director of the Radioactive Waste Management Section within the Department of Industry, Science and Resources, visited Coober Pedy to conduct a regional consultation session.

The Kungkas refused to see the pro-dump video, insisting on a private session with the Project Director and her assistant. They employed their own interpreter, Pauline Lewis, and for two hours spoke strongly about their *Tjukur*—Dreaming and their responsibilities for the country.

Written soon after hearing about the waste dump proposal for the Billa Kalina region, this statement circulated around Australia and the world.

> We are the Aboriginal Women.
> Yankunytjatjara, Antikarinya and Kokatha.
> We know the country.
> The poison the Government is talking about will poison the land.
> We say, "No radioactive dump in our ngura—in our country."
> It's strictly poison, we don't want it.

We were born on the earth, not in the hospital. We were born in the sand. Mother never put us in the water and washed us when we were born straight out. They dried us with the sand. Then they put us, newborn baby, fireside, no blankets, they put us in the warm sand. And after that, when the cord comes off, they put us through the smoke. We really know the land. From a baby we grow up on the land.

Never mind our country is the desert, that's where we belong. And we love where we belong, the whole land. We know the stories for the land. The Seven Sisters travelled right across, in the beginning. They formed the land. It's very important *Tjukur*—Law, the Dreaming, that must not be disturbed. The Seven Sisters are everywhere. We can give the evidence for what we say, we can show you the dance of the Seven Sisters.

Listen to us! The desert lands are not as dry as you think! Can't the Government plainly see there is water here? Nothing can live without water. There's a big underground river underneath. We know the poison from the radioactive dump will go down under the ground and leak into the water. We drink from this water. Only the Government and people like that have tanks. The animals drink from this water: *malu*—kangaroo, *kalaya*—emu, porcupine, *ngintaka*—perentie lizard, goanna and all the others. We eat these animals, that's our meat. We're worried that any of these animals will become poisoned and we'll become poisoned in our turn.

The poison the Government is talking about is from Sydney. We say send it back to Sydney. We don't want it! Are they trying to kill us? We're a human being. We're not an animal. We're not a dog. In the old days the white man used to put poison in the meat, throw them to feed the dogs and they got poisoned, straight out and then they died. Now they want to put the poison in the ground. We want our life.

At the Breakaways. Eileen Crombie, Eileen Brown, Dianne Edwards, Eileen Wingfield, Renee Wintinna, Emily Austin.

"Never mind our country is the desert, that's where we belong"

All of us were living when the Government used the country for the bomb. Some were living at Twelve Mile, just out of Coober Pedy. The smoke was funny and everything looked hazy. Everybody got sick. Other people were at Mabel Creek and many people got sick. Some people were living at Walatina. Other people got moved away. Whitefellas and all got sick. When we were young, no woman got breast cancer or any other kind of cancer. Cancer was unheard of with men either. And no asthma. We were people without sickness.

The Government thought they knew what they were doing then. Now, again they are coming along and telling us poor blackfellas, "Oh, there's nothing that's going to happen, nothing is going to kill you." And that will still happen like that bomb over there.

And we're worrying for our kids. We've got a lot of kids growing up on the country and still coming more, grandchildren and great grandchildren. They have to have their life.

It's from our grandmothers and our grandfathers that we've learned about the land. This learning isn't written on paper as whitefellas' knowledge is. We carry it instead in our heads and we're talking from our hearts, for the land. You fellas, whitefellas, put us in the back all the time, like we've got no language for the land. But we've got the story for the land.

Listen to us!

Ivy Makinti Stewart, Eileen Kampakuta Brown, Eileen Unkari Crombie, Eileen Wani Wingfield, Lucy Kampakuta Wilton, Pingkai Upitja, Emily Munyungka Austin, Angelina Wonga, Peggy Tjingila Cullinan, Dianne Edwards.

Kupa Piti Kungka Tjuta

The main street.

Committee members hold a stall.

Sister Michele Madigan gears up for the Opal Festival.

Coober Pedy gets active

March

A groundswell of local activity against the dump proposal occurred in Coober Pedy, outback opal mining town 850 kms north of Adelaide.

Sharon Drage, a local working at the post office, organised a public meeting about the dump. Over 40 people attended and formed the Coober Pedy Committee Against the Radioactive Waste Repository.

Committee members were extremely creative and active over the years that followed. They organised numerous petition campaigns, countered Government misinformation, networked with anti-dump groups, protested at local Opal Festivals and held public meetings.

The Committee applied concerted pressure on the District Council to take a stance. This resulted in Coober Pedy being declared a Nuclear Free Zone in early 2000 after hundreds of locals signed the petition.

Sister Michele Madigan, the Kungkas' 'honorary paperworker' was actively involved in the Coober Pedy Committee, and the two groups often worked together over the coming years.

Letter to Dr Caroline Perkins, Department of Industry, Science and Resources, following the regional consultation session held in Coober Pedy in April.

June 5, 1998

To Caroline Perkins,

We women, traditional women here in Coober Pedy, made a strong effort to come to your day here in Coober Pedy about the radioactive dump. We spoke to you for two hours and tried to educate you on many things you don't know about, including the land, the water present in the desert, that you don't understand. We also explained to you about the *Tjukur*—Dreaming present in the area.

You promised that you would get in touch with us, to show you listened and understood. But it's just like our words went in the wind.

Now we hear the place is close to Coober Pedy, 150 kms. It's too close. Really close. *Kapi wiya*, the water will be affected. You have to come back here and talk to both man and woman.

Eileen Wani Wingfield and Eileen Unkari Crombie

Kupa Piti Kungka Tjuta

"It's just like our words went in the wind"

Inma at Ten Mile Creek Elders Camp.

Letter to Friends of the Earth (FoE) Melbourne after Eileen Crombie suggested, "Nobody is listening to us about the dump, get the greenies."

August 3, 1998

Dear Greenies,

We are just dropping a few lines to let you know we want help. We're trying hard about this rubbish, the radioactive dump. We don't want that, we've got kids, we've got too many kids to grow up and see the country. We've been talking to the Government, but they've taken no notice.

We've got underground water, that's why we're worrying about the water. It would be different if we had water coming from a pipeline somewhere, but we depend on the underground water. We don't want the poison from the dump leaking into underground water.

But anyway, we don't want the dump, the poison. Send it back to where it came from. They've got an *apu*—hill there, the Blue Mountains. Dig a hole there and put cement in it. The ground is too soft up here. Or put it in Canberra, John Howard's backyard.

And we're worried about the animals. We eat kangaroo, emu, (they've already poisoned the rabbit), *ungkata*—lizard, goanna, cadney, *ngintaka*—perentie, porcupine, *kipara*—wild turkey. We worried about all our beautiful native birds. We worried that any of these animals and birds will become poisoned and we'll become poisoned in turn.

We're worried that the poison will seep down into the river under the ground and poison our *maku* and *tjala*—witchetty grubs and honey ants. These are our bush tucker, the bush tucker of our grandmothers and grandfathers; we're using all these things, eating wild banana, wild plum, wild tomato, quandong and also *Irmangka Irmangka*—the bush medicine plant.

We want help! We want you to come up here to Coober Pedy and have a meeting with the Aborigine people, and any whitefellas from here who want to come. We'll invite all the locals. We want you please to send us back a fax to let us know how you can help us.

There are people getting sick in Roxby. We know that ourselves. Some are leaving the job. We know the dump people and the Government, the ones giving orders, are sitting down clean over there.

Eileen Unkari Crombie, Peggy Tjingila Cullinan, Ivy Makinti Stewart, Eileen Wani Wingfield, Eileen Kampakuta Brown, Angelina Wonga, Emily Munyungka Austin, Betty Nyangala Muffler.

Kupa Piti Kungka Tjuta

Emily Austin and Eileen Crombie prepare Irmangka Irmangka—traditional bush medicine.

"We don't want the poison from the dump leaking into underground water"

The Kungkas travelled to Melbourne to attend the Global Survival and Indigenous Rights conference, hosted by FoE Melbourne.

Women united by a really hot topic

By Carolyn Webb
The Age
November 21, 1998

It was a sunny 21 degrees in Maribyrnong yesterday, but a cold day for the Kupa Piti Kungka Tjuta women's group. Back home it gets up to 50 degrees in summer. But the 16 women from Kokatha, Arabunna and Antikarinya Aboriginal people, communities from Port Augusta to Coober Pedy in South Australia, were not here to talk about the weather.

The Federal Government wants to put a national radioactive waste repository on Kokatha land, 180 kms south-east of Coober Pedy. The women voiced their opposition to the proposal at the opening of the second annual Indigenous Solidarity Gathering at Pipemakers Park. The waste comes from all over Australia and authorities want to dump it in the desert, where the women say it will infiltrate nearby Aboriginal communities' air, bore water and land.

A spokeswoman, Ms Rebecca Bear-Wingfield, said most of the women, "my mothers, my grandmothers, my aunties and my sisters", rarely left their own communities. "Some people had a six hour drive from Coober Pedy to Port Augusta where we left at 5 am on Tuesday on the first Ghan train to Melbourne and we got here at 10 pm the next night," she said.

"But the fact that so many women from such a diverse area are prepared to come to Melbourne highlights the importance of this issue for our people." She said it was highly dangerous to move thousands of barrels of nuclear waste to a place that had already housed the Woomera rocket range and the Nurrunga American Spy Base.

Ms Wingfield's mother was exposed to radiation from atomic testing at Emu Junction near Coober Pedy in 1953. Some women present yesterday had walked in craters from the Maralinga blast as children.

Ms Bear-Wingfield said: "We're just like any other people. We're family who are concerned for our family and our children's children."

On the banks of the Maribyrnong River. Top: Betty Muffler, Eileen Crombie, Lucy Wilton, Alma Allen, Betty Amos. Middle: Emily Austin, Lallie Lennon, Eileen Brown, Craig Cullinan, Eileen Wingfield, Lola Amos. Bottom: June Lennon, Rebecca Bear-Wingfield, Joanne Gaston.

DECEMBER 1

GLOBAL WASTE DUMP PLOT EXPOSED

A promotional video by multinational nuclear waste management company, Pangea Resources, is leaked to the Australian media. The video recommends the Australian outback as 'ideal' for permanent disposal of the global stockpiles of high-level nuclear waste.

"Our plan is a global solution to a global problem and it is our fervent wish to make a significant contribution to world security." Pangea Resources.

Open letter after hearing about the leaked Pangea proposal for an international nuclear waste dump in the Australian outback.

December 16, 1998

Re: the idea of being the dump for the whole world

They really want to kill us!

They really are aiming to wipe the country out, not just us but all the living things in the whole earth.

They might as well come and kill us straight out. Kill us like a dog in the days long ago instead of this sneaky way of killing us. Kill us straight out.

A long time ago they been brainwashing the Old People about the land, brainwashing them with tobacco and rations, and they've been running them around like sheep. But some other Old People sat down and talked to us about the land. At night-time and in the daytime, before daybreak as a morning message—*aalpiri*. It's from these grandmothers and grandfathers that we've learned about the land. This learning isn't in our heads and we're talking from our hearts, for the land.

You fellas, whitefellas, put us in the back all the time, like we have no language for the land. But we've got the story of the land. We're humans the same as you, only the colour is different.

Please help us!

Ivy Makinti Stewart, Eileen Wani Wingfield, Eileen Kampakuta Brown, Angelina Wonga, Eileen Unkari Crombie, Emily Munyungka Austin, Lucy Kampakuta Wilton.

Kupa Piti Kungka Tjuta

"We've got the story of the land"

Emily Austin, Martha Edwards, Betty Muffler, Eileen Wingfield, Ivy Stewart.

1999

Emily Austin and Eileen Crombie with MKs at Bon Bon Station.

Melbourne Kungkas

January

The Kungkas' visit to Melbourne in 1998 inspired a group of young non-Aboriginal women to respond to their request for support.

The Kungkas named the group Melbourne Kungkas (MKs) as they began corresponding with Coober Pedy via fax. The MKs started by raising awareness and campaign funds, selling postcards, holding raffles and organising benefit gigs.

Initially, the campaign relied heavily on the resources of the MKs and others in the city and over time, the MKs came to play a crucial role in building the national profile and support base of Irati Wanti. The MKs helped generate media and link up Irati Wanti with other greenie groups and issues. Through visits, phone calls and emails the MKs worked closely with the Irati Wanti coordinators; they provided vital friendships as well as logistical support.

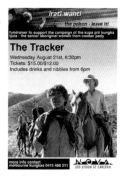

Melbourne Kungkas 1999-2004 included:

Lucy Brown, Nina Brown, Kathlyn Brown, Amelia Young, Nina Cunningham, Camilla Pandolfini, Samantha Sowerwine, Clare Brown, Georgina Wright, Corinne Balaam, Breony Carbines, Sophie Bourke, Eve Vincent, Shannon Owen, Clare Land, Rafaela Pandolfini, Meg Jones, Christie Hannan, Rachel Barnden, Sarah McCall, Jen Nicita, Aimee Pilven, Jeanie Tsang, Linda Odgers, Fiona Hallam.

Opal Festival

April

"We're doing our best."

"Not everything buried is dead."

The Kungkas were joined by the Coober Pedy Committee and visiting Melbourne greenies in Coober Pedy's annual Easter street parade.

The Kungkas travelled up the dusty main street in their Toyota Troupe Carrier, proudly holding placards and singing *Inma* out the windows. Local *Anangu* families prepared colourful banners and placards with strong words about looking after the country. They were awarded Best Walking Float.

MAY 8

SA UNDER FIRE

The Federal Government pressures SA to accept the dump in spite of recent news polls revealing community opposition as high as 95%.

"The dump proposal represents logical, well-protected use of an isolated part of Australia... SA would be making a contribution to the nation. If we don't do this, it will send another message that the place is closed for business." Nick Minchin, Federal Minister for Industry, Science and Resources (and SA Senator).

Radioactive Exposure Tour

June

FoE Melbourne organised a tour from the Lucas Heights nuclear reactor in Sydney to outback SA. Fifty greenies embarked on a 10 day journey to expose the massive expansion planned for Australia's nuclear industry.

From Lucas Heights, the production site for the majority of waste destined for the planned dump, they travelled the proposed waste transport route through Western NSW. The tour stopped at communities along the Barrier Highway meeting with local people, the media and councils.

After touring Honeymoon and Beverly uranium mines, the exhausted group finally arrived at Ten Mile Creek Elders Camp, near Coober Pedy. The Kungkas welcomed the travellers and sang *Inma* into the night.

At a meeting the next day, the Kungkas first flagged the idea of greenie *kungkas*—women moving up to Coober Pedy to "open an office... help with stopping the poison".

From the city to the outback.

Humps not Dumps

June-August

Eight women trekked 1,000 kms across Billa Kalina in a spirited bid to draw attention to nuclear activities in SA.

In support of the Kungkas' fight Luna, Wren, Julia, Janine, Mel, Izzy, Sophia and Catherine braved harsh desert conditions for three months on their anti-nuclear camel trek.

After training wild camels with Coober Pedy local Phil Gee, the cameleers met with the Kungkas at Warina Siding for the trek's launch. The Kungkas shared fond memories of travelling around on camels when they worked on stations as young women.

Humps not Dumps travelled from Warina Siding, along the Oodnadatta Track to Olympic Dam uranium mine at Roxby Downs. From there they journeyed through Woomera and then to Maree, Copley and Leigh Creek.

Mel Stron, Wren Redback, Izzy Brown.

Talking to school kids along the way.

Humps not Dumps continued...

A few months later the Kungkas met with Humps not Dumps at Olympic Dam, performing *Inma* at the gates of the world's third largest uranium mine.

Humps not Dumps attracted national and international publicity, contributing to growing anti-waste dump sentiment in South Australia.

Cameleers.

Eileen Brown.

Movie star at Ten Mile

September

In Coober Pedy to film Hollywood flick *Red Planet*, Val Kilmer asked to visit the Kungkas and their families at Ten Mile Creek Elders Camp. Sitting around a campfire, the Kungkas told him about the *Tjukur*—Dreaming, and their worries about the waste dump.

This brush with stardom led to articles and photos in celebrity magazine *Who* and the *Adelaide Advertiser*.

A brush with stardom. Rebecca Bear-Wingfield, Lucy Wilton, Eileen Wingfield, Val Kilmer, Eileen Crombie.

Kungkas sing to a full house

November

A public meeting organised by the Australian Conservation Foundation (ACF) was held at the Adelaide Town Hall to activate community support for the anti-dump campaign.

The Kungkas convoyed to Adelaide in three cars to attend the meeting. As they drove through the night the van in which Ivy Stewart, Eileen Brown, Eileen Crombie and Angelina Wonga were travelling, hit a bullock on the stretch of highway between Pimba and Port Augusta. Luckily no one was seriously injured, confirmed by a check up at the doctors in Port Augusta the next day. Despite the shock of the car accident the Kungkas insisted on travelling through to Adelaide for the public meeting. "We have to keep going, we have to sing *Inma*."

Community members gather at Adelaide Town Hall.

Eileen Brown and Lucy Wilton sing Inma. "That's exactly what the song says; I left it [irati—poison] in the ground, you leave it in the ground."

Punch Gibson talks to the full house through interpreter, Mona Tur.

Punch Gibson, Eileen Crombie, Angelina Scobie, Eileen Wingfield, Angelina Wonga, Mona Tur, Ivy Stewart.

Dump in SA? No way.

Full house continued...

Almost 1,000 people filled the seats of the Adelaide Town Hall, greeted by Ivy Stewart's extraordinary singing. As MC David Noonan recalls, "Here was the capacity of the South Australian movement against the dump and the Federal Government were shivering in their boots."

Speakers included Rebecca Bear-Wingfield (Kupa Piti Kungka Tjuta), Lyn Allison (Democrats), John Hill (SA Shadow Environment Minister), Peter Garrett (ACF) and Dr Jim Green (Lucas Heights campaigner).

Eileen Wingfield's daughter Rebecca Bear-Wingfield addressed the meeting regarding the human rights and cultural obligations of the Kungkas and their communities.

The Kungkas then put forward Punch Miani Gibson, Eileen Crombie's uncle, to speak for them.

It was the first time the Kungkas appeared in front of such a large crowd and Eileen Brown and Lucy Wilton, whose tee shirts were covered in white and yellow paint to symbolise the *irati*—poison, overcame their stage fright and walked onto the stage. The crowd was silent as they performed *Inma* about the *irati*. A standing ovation followed.

The diversity and strength of the community revealed at the public meeting wasn't ignored. Extensive media coverage and political debate followed. The next day the Liberal Premier John Olsen split from the Federal Liberal Party position by publicly opposing any high-level waste dump for SA.

2000

Letter to Dorothy Kotz, SA Minister for the Environment and Aboriginal Affairs (Liberal).

February 17, 2000

Dear Dorothy Kotz,

Pinangku kulila—please listen to us.

We want to see our kids grow up. We don't want our kids to be poisoned. We want our life.

Every time we're talking our words bounce back. Over and over we're saying the same thing. Don't let them bring the poison from where they are treating it in Sydney. Put it back there in Sydney. It's too dangerous, the trucks bringing it all the way here.

Yes, over and over, we're saying the same thing. It's just like we're bouncing a ball and nobody's catching it. *Manta winki*—it's the whole country that's got the *Tjukur*—Law: North, East, South, West.

You're a woman like us. *Anangu kulila*—listen to us Aboriginal ladies. Don't be leading us around like a camel. Don't put us in the back.

You're welcome to come up here to Coober Pedy. We feel sorry that you don't know much about the land, about the culture. Come up here to Ten Mile Creek and we'll help you listen about the culture. It will help you understand about culture and the land together. We're inviting you.

Please listen to us!

From all the Ladies,

Kupa Piti Kungka Tjuta

"We don't want our kids to be poisoned. We want our life."

Vincent Warren, Ringo Wells, Milika Crombie, Isaac Warren.

Irati Wanti headquarters

Wilson Road office.

March

In 1999 the Kungkas had suggested that some greenie *kungkas*—women move from the city to Coober Pedy to establish a campaign office. In March 2000, Lucy Brown arrived from Melbourne in her bright yellow Bedford van.

Lucy set up a humpy out at Ten Mile Creek Elders Camp. She scavenged some essentials from the tip and got to work on the campaign in the storeroom out the back of the Kungka Tjuta office on the main street.

At an early meeting Kungka Lucy Wilton suggested calling the office "*Irati Wanti*—the poison, leave it". The other Kungkas agreed, and so the campaign was named. Slowly, the Irati Wanti campaign 'office' became a desk and computer in a dugout (underground house). Eventually it would become a busy, fully functioning office in Bienke Street on the outskirts of town.

In its first few years the office focused on building up a national contact list of supporters, networking with the MKs and other anti-dump groups, establishing the campaign's identity, trying to keep the computer free of dust, and of course, "getting the [Kungkas'] words out," as Eileen Crombie often said.

The demands on the office increased as the campaign grew. By 2003 the office was the central contact point for extensive media and political liaison.

"We look after the country together"

"We can talk strong, straight out, but we can't do anything without the pen and paper and that's the whitefella way. That's where the greenie *kungkas*—women came in. Learning and teaching: we learn from you, you learn from us. We come together, us ladies and the greenie girls, all the way, hand in hand we travel along. The greenie girls did a good job to help us and we teach them back, A̱nangu way. Hand in hand for the future, that's the way. We look after the country together."

Emily Austin and Myra Watson.

Over the years six greenie *kungkas* made tracks from down south to work with the Kungkas on the campaign.

Lucy Brown

Nina Brown

Emily Johnston

Clare Brown

Georgina Wright

Alexandra Kelly

Opal Festival

April

Melbourne Kungkas were in town for the annual Opal Festival.

The Kungkas teamed up with the Coober Pedy Committee to produce a spectacular parade entry. Local opal miners constructed mock waste storage tanks that were fixed to the back of a big truck.

Kungkas in the capital

May

Eileen Wingfield, Eileen Brown, Lucy Wilton, Peggy Cullinan, Maureen Williams and Sister Michele Madigan travelled to Canberra for a book launch. The children's book *Down the Hole, Across the Sandhill, Up the Tree: Running from the State and Daisy Bates* was written by Eileen Wani Wingfield and Edna Tatingu Williams and illustrated by Kunyi June-Anne McInerney. Democrats Senator Aden Ridgeway launched *Down the Hole* at Parliament House.

The Kungkas also visited the Aboriginal Tent Embassy, met with Lyn Allison (Democrats) and Nick Bolkus (ALP Shadow Environment Minister), and spoke to national media outside Parliament House.

Their message was loud and clear. "We've been fighting against this radioactive waste, this poison, for more than two years. Children are being born and growing up and still we are talking against the poison, the radioactive waste."

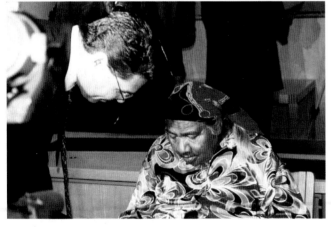

Eileen Wingfield with Senator Aden Ridgeway.

Outside Parliament House.

Walking for peace

May

The Kungkas travelled to Lake Eyre South to help launch Walking the Land - Walk for Peace.

Keeper of Lake Eyre, Arabunna Elder Kevin Buzzacott said, "We are walking our land, in the company of the Old People and the old spirits of the ancient land. We are walking from Lake Eyre to Sydney, to arrive before the Olympic Games, carrying the real flame with us, the sacred firestick with our big message of peace to the world."

The Kungkas sang *Inma* as Kevin and his supporters left Lake Eyre's shores, embarking on their four month journey overland to Sydney via Canberra. The Kungkas started planning to meet the walkers at the Sydney Olympics.

"My brother was very happy we came to support him looking after the country" Lucy Kampakuta Wilton

Lucy Wilton presents Kevin Buzzacott with beads, Betty Muffler looks on.

Letter to Kim Beazley, Leader of the Opposition, before the ALP National Conference.

July 27, 2000

Dear Mr Beazley,

We think you're a good leader. But we're writing against the radioactive dump. We want everyone to be against it and we're asking for your help against it too. The meeting is your chance to stop and think. Don't be like John Howard messing up the country, our beautiful country. We've got grandkids and you've got grandkids too, all the whitefellas got grandkids. That's what we're worried about, our future.

We've got water in the desert, underground water. No pipes from the south up here. Animals like kangaroo and emu are our meat and even your fellas' meat, sheep and cattle, drink the water that comes from the underground. Poison, radioactive waste, leaks into the water. So everybody should think and help, for our future, without argument, because we've got a beautiful country, Australia and everyone comes and looks for peace in Australia and there's plenty of room for everyone.

We're here to look after the country. We're not going to live forever. If we do the right thing to help the younger generation, they'll turn around and fight for the protection of their country in their turn. And many tourists come from all over to our country, to see the scenery and enjoy the sunshine, no poison, nothing. We don't want nothing, no radioactive dump buried in Australia.

Emily Munyungka Austin

Kupa Piti Kungka Tjuta

"That's what we're worried about, our future"

Emily Austin with her grandkids; Isaac Warren, Robert Austin, Cherika Lang and Emma Austin.

Race Around Oz

August

Ivy Stewart, Eileen Brown, Eileen Crombie and Angelina Wonga featured on ABC TV's popular competition series *Race Around Oz*.

Dahlia Abdel-Aziz's five minute story, *Following The Sisters' Journey*, showed the Kungkas' strength in their culture, and in surviving the atomic tests at Emu Junction and Maralinga.

Dahlia had previously worked with the Kungkas as an anthropologist. She won the judges' prize in the competition.

Film-maker Dahlia Abdel-Aziz.

Olympic vision of the Seven Sisters

By Lucy Brown
Coober Pedy Regional Times
October 4, 2000

Coober Pedy Elders visit Sydney's Olympic Games with a message of peace and strength of culture.

The senior Yankunytjatjara women undertook a journey of 3,000 kms by train (a journey which took two days) to be a part of the international media spotlight surrounding the Olympic Games.

"We were brave ladies, sitting right at the door of Lucas Heights." Eva Williams, Emily Austin, Dianne Edwards, Martha Edwards, Eileen Crombie, Eileen Brown, Myra Watson.

Olympic vision continued...

The Aboriginal Tent Embassy in Victoria Park during the Olympics.

Captain Cook's Foot, Botany Bay.

Eileen Crombie and Eileen Brown perform Inma on the shores of Botany Bay.

Next page: Angelina Wonga at Victoria Park.

"We are *ninti pulka*—very knowledgeable women for *Tjukur*—for the culture. We are taking the beautiful desert culture to Sydney and there we will dance the culture and sing the story of the Seven Sisters," said the Kupa Piti Kungka Tjuta.

The Kungka Tjuta's concern for their country was highlighted by their visit to the Lucas Heights reactor, the place of production for the vast majority of Australia's nuclear waste.

"We were brave ladies, sitting right at the door of Lucas Heights. Going to talk about our country. We sat right around talking about the radiation dump. We don't want it," said Emily Austin.

At the reactor, the Kungkas held a very productive meeting with environmentalists and councillors of the Sutherland Shire Council who are campaigning to stop the construction of a $300 million reactor, proposed for the same site.

"The A_nangu people know [to leave the uranium lying in the ground]," said Eileen Crombie. "Always."

In a busy seven days in Sydney, the Kungka Tjuta found time to perform their *Inma*—dances, stories and creation songs for the land, both at the Aboriginal Tent Embassy in Victoria Park, Central Sydney and at their camping place with Arabunna Elder Uncle Kevin Buzzacott, at 'Captain Cook's Foot'. Their camping place at 'The Foot', as it is named by Mr Buzzacott, is the place where Captain Cook first landed in Australia and is the destination of the Walking the Land - Walk for Peace.

After awaiting the arrival of the Prime Minister John Howard, who had been invited but unfortunately did not attend, the Kungkas, Mr Buzzacott and the walkers held a ceremony for peace and for the healing of the land.

Mr Buzzacott and Kungkas Dianne Edwards and Eva Williams took a dawn boat trip to the Prime Minister's residence, Kirribilli House, and around Sydney Harbour, cleansing the area with smoke from the Lake Eyre sacred fire.

The Kungka Tjuta extended an invitation for all whitefellas to come and sit down with them and to learn about their culture.

"We are inviting all the whitefellas to come and sit down with us by the fire and listen and learn the Aboriginal way," said Emily Austin. "We want the whitefellas to understand about A_nangu culture. The culture isn't dying, it's still alive. This land is our country for so long. And it's for them too, eh? By listening to us talking for our country, you're helping to save our land, our Australia."

36 Talking Straight Out

Lucy Wilton, Eileen Wingfield, Emily Austin.

ABC journalist Nance Haxton interviews Emily Austin.

Kungkas are Great

November

Eileen Brown, Emily Austin, Eileen Wingfield and Lucy Wilton proudly received a South Australian Great Regional Award at a presentation evening in Whyalla.

In the regional award's first year, the Kungkas were honoured in the Environment and Health category for "their actions and ideas, that ultimately contribute to the building of a more positive and vibrant region".

On the way to Whyalla for the award ceremony, the Kungkas stopped off in Port Augusta for a meeting with members of local group CARDS (Campaign Against Radioactive Dumps).

Returning to Coober Pedy the Kungkas celebrated with younger group member Dianne Edwards, and office worker Renee Wintinna.

Lydia Smith, Lucy Wilton, Eileen Wingfield, Eileen Brown, Emily Austin, Leonie Bakk, Mary Cusak.

Dianne Edwards.

Renee Wintinna.

Irati Wanti online

November

The campaign website was constructed with the generous technical and financial support of Sam de Silva and *myspinach.org*. Melbourne website producer Adam Grubb spent two weeks in Coober Pedy working with the Kungkas and site curator Nina Brown.

Adam designed the site's banner, which became the campaign's key image. The site was set up with open publishing functions, which meant the Irati Wanti coordinators could update its news, event and photo journal sections from Coober Pedy.

Officially launched in early 2001, *www.iratiwanti.org* was a lifeline for the campaign. Adam continued on as Irati Wanti's treasured webkeeper, upgrading the site on an unpaid basis until 2004.

Webkeeper Adam Grubb.

NOVEMBER 16

LIBERALS SPLIT ON DUMP

The SA Liberal Government splits from the Federal Liberal Party position on the dump. They pass legislation that bans the importation and dumping of intermediate and high-level radioactive waste, but fail to legislate against the low-level dump.

"We are sending a clear message to Canberra that such a dump is totally unacceptable." Iain Evans, SA Environment Minister (Liberal).

Irati Wanti, the film

December

Irati Wanti, a 17 minute documentary film directed by MK Shannon Owen, premiered at Wild Spaces Film Festival in Melbourne.

Shannon was committed to producing a timeless, awareness raising film. It became a vital campaign resource, screening to national and international audiences in schools, universities, public meetings, conferences and film festivals.

Film-maker Shannon Owen.

A travelling crusade - Elders win award for dump site campaign

By Thea Williams
Adelaide Advertiser
December 16, 2000

It was another long bus drive overnight from Coober Pedy for the five women from the senior Aboriginal women's council, the Kupa Piti Kungka Tjuta, to Adelaide yesterday.

But long drives are nothing new to the group of Elders from the Arabunna, Kokatha, Antikarinya and Yankunytjatjara communities.

The group has spent two years patiently but persistently travelling the country in its crusade against the Federal Government's plans for a nuclear waste dump in SA, including trips to Canberra, Sydney and Melbourne. The Elders from the group travelled the 850 kms to Adelaide to receive the $1000 Jill Hudson Award for Environmental Protection from the Conservation Council of SA.

They have a simple message to send "Irati Wanti (the poison, leave it)". "Radiation will get into the underground water," Emily Munyungka Austin said yesterday. "We're not worried about money, we're worried about poison for our kids, we're worried about life."

Three sites around Woomera and Roxby have been suggested by the Federal Government for the location of a low-level waste dump. A decision is expected to be made next year.

The women live under the cloud of British nuclear tests in the 1950s at Emu Junction. "It went up close and the next morning I woke up red eyes and coughing. You could see it as all light," Eileen Kampakuta Brown said.

While driving thousands of kilometres to save their heritage, the women have also become champions of the internet.

Statements from the group appear on the Australian Conservation Foundation website, the National Women's Network website and even in chat rooms on websites in French, German and Japanese, not to mention their own website under construction: *www.iratiwanti.org*.

Five of the eight women who make up the council made the trip from Coober Pedy: Emily Munyungka Austin, Eileen Kampakuta Brown, Eileen Unkari Crombie, Angelina Wonga and Lucy Kampakuta Wilton.

The award was set up in the name of Jill Hudson, a South Australian primary school teacher with a passion for teaching young people about the environment. Her husband, Dr Dennis Matthews, said yesterday: "We want to encourage people to be whistle blowers not the run-of-the-mill greenies."

JANUARY 25

COMMONWEALTH LAND CHOSEN FOR DUMP

The preferred site for the dump is announced for Commonwealth Defence land in the Woomera Prohibited Area. "Site 52a at Evetts Field West has been selected as the preferred site following rigorous scientific assessment and taking account of extensive consultations with regional stakeholders." Nick Minchin, Federal Minister from Industry, Science and Industry.

Nuclear waste fight on the web

Adelaide Advertiser
January 26, 2001

Aboriginal Elders yesterday embraced dot.com technology as part of their attempt to rally opposition to a federal plan to make South Australia the national dumping ground for low-level radioactive waste.

Kupa Piti Kungka Tjuta women in the state's outback branded the Howard Government's community consultation process a sham.

Their website, *www.iratiwanti.org* which means "the poison, leave it", chronicles a campaign against the nuclear industry.

On Wednesday, Industry, Science and Resources Minister Nick Minchin, named a desert plateau north-west of Woomera as the Government's preferred location for a low-level nuclear dump.

The Aboriginal Elders yesterday complained they had been excluded from negotiations relating to the dump. "Australia talks of reconciliation, but how can we reconcile when this waste is going to be dumped on the ancestral lands of the Kokatha people," Aboriginal woman Rebecca Bear-Wingfield said.

Kalta—sleepy lizard.

FEBRUARY 8

CO-LOCATION RULED OUT

Federal Government buckles under community pressure and rules out co-location of an above ground store for intermediate and high-level waste.

"It's time to put an end to the deliberate scare campaign by extremist groups seeking to whip up community concern about co-location." Nick Minchin, Federal Minister from Industry, Science and Industry.

The low-level waste dump is still planned as an 'out of sight, out of mind' solution to material that remains radioactive for up to 300 years. The dump's design has not been officially released, but it is thought to be the size of a soccer field, with radioactive waste stored in steel drums in shallow unlined trenches and covered with rammed earth or concrete.

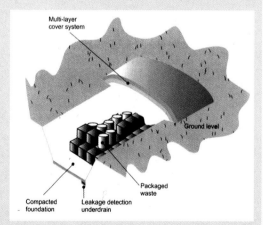

Proposed design of the low-level waste dump.

Letter to Natasha Stott Despoja, recently elected leader of the Australian Democrats.

April 9, 2001

Dear Natasha,

Congratulations for becoming leader of the Democrats. We are the Kupa Piti Kungka Tjuta - the Senior Aboriginal culture women from Coober Pedy, South Australia.

We're writing to let you know that you gotta talk up for us in Government. We've been fighting against the uranium for a long time. Roxby first and now this other thing that they want to bring back and bury. The nuclear waste dump. And we've been through the bomb (Emu Junction, Maralinga).

Then they're starting a new mine up here in Lake Phillipson. *Pina pati*—their ears are blocked. This country is full of sites. Why are the men making all the decisions? It's all blokes doing this dump. The mining. The bomb. Nobody asked us! We're against it all!

They gotta listen to the *kungkas*—women. *Kungkas* are still the boss. *Kungkas* are still mother nature. They come from the *kungkas* and they grow up from the *kungkas' mimi*—breast. They gotta think about being equal. Not just the men. We're thinking about the kids and the future generation.

We are very busy women working to look after country all the time. We've been working very hard. It's very important you know that.

We invite you to come here and sit with us and learn.

Ivy Makinti Stewart, Eileen Kampakuta Brown, Eileen Wani Wingfield, Emily Munyungka Austin, Eileen Unkari Crombie.

Kupa Piti Kungka Tjuta

At Lake Phillipson. Betty Muffler, Edie King, Eileen Brown.

"We are very busy women working to look after country all the time"

Letter to Jawoyn people facing uranium mining exploration in Arnhem Land, Northern Territory.

21 April, 2001

Dear Jawoyn People,

Hello. We are the Kupa Piti Kungka Tjuta. We are the Senior Aboriginal cultural women in Coober Pedy, South Australia. We are dropping you a line to let you know about our struggle against the uranium and how we are standing up strong for country. We are writing to let you know that uranium is dangerous and not to give them the right of way.

Everybody's been under the bomb here when they set them off here in the fifties. All the Old People died and all the families are sick. First there was Emu Junction, two tests, which they covered up in October 1953. And then there was Maralinga which you might of heard of? We can't go back there ever. Those places are messed up forever.

There used to big shady trees here, all over, all the way down. In the big rain time, the tablelands here used to be covered with purple flowers. Now it is all dry, dry as a bone, and terrible. It has been poisoned. We used to be healthy people. We used to walk everywhere. All our kids are sick now. You don't get one healthy kid nowadays.

And now we are fighting against the nuclear dump. And we don't want that. We don't want more poison just like the bomb. We will attach another paper so you can read all about it and we can also send you a copy of our video how we are talking about it, against the poison. And you can look at out website *www.iratiwanti.org*.

Don't give them any OK! Any mining company to mine the uranium.

SAY NO. N - O!

We are fighting for it to stay in the ground. It's dangerous. Very dangerous. It's not the money we are worrying about. We are worrying about life.

Eileen Wani Wingfield, Eileen Kampakuta Brown, Ivy Makinti Stewart, Emily Munyungka Austin, Angelina Wonga, Tjunmutja Myra Watson, Dianne Edwards, Fanny O'Toole, Jeanne Winter, Rebecca Bear-Wingfield, Renee Wintinna.

Kupa Piti Kungka Tjuta

Craig Cullinan and Eileen Brown.

"We are fighting for it to stay in the ground"

MKs solidarity road trip

July

Eight Melbourne Kungkas spent a week in Coober Pedy for an office working bee, to meet with the Kungkas and to get involved in community activities. They screened films at the local TAFE and held an information stall in the main street, selling *Irmangka Irmangka*—traditional bush medicine. The MKs also ran circus workshops for kids at Umoona Community and the Area School and cooked up a feast for the Kungkas and family members.

Eve Vincent and Camilla Pandolfini at the stall outside Coober Pedy TAFE.

Circus workshops at Umoona basketball courts. Top: Marissa Buzzacott, Katie Gibson, Tanita Van Horen. Bottom: Jen Nicita, Shannon Owen, Meg Jones, Eve Vincent.

Naomi Klein in town

July

The best selling Canadian author and her partner, Avi Lewis, visited Coober Pedy before a hectic Australian speaking tour. Naomi's book *No Logo* explores the rise of corporate culture and globalisation, and charts ongoing resistance by local communities.

Naomi and Avi came to visit the Kungkas and hear their stories firsthand. Unfortunately they were unable to meet with the Kungkas due to Sorry Business (grieving time after an *A̲nangu* community member passes away). Instead, they sat around a campfire with the Irati Wanti coordinators and visiting greenies. Naomi and Avi became long-term Irati Wanti supporters.

Outside Wilson Road office. Avi Lewis, Nina Brown, Lucy Brown, Aren Aizura, Heidi Douglas, Naomi Klein.

Don't waste our country

September

Irati Wanti and Alliance Against Uranium released the Forever Country poster. Forever Country is a project of Alliance Against Uranium, a network of Aboriginal and non-Aboriginal people opposed to uranium mining and the nuclear industry.

Lisa Robins designed the poster, based on web designer Adam Grubb's image. The poster was distributed across Australia and generated much publicity about the campaign.

Yeperenye Festival

September

Thousands of Aboriginal and Torres Strait Islander people converged in Alice Springs for the Yeperenye Indigenous Federation Festival. Celebrating a "Federation of Indigenous Nations" the Kungkas joined with over 4,000 Indigenous performers from 30 nations to sing and dance stories from their respective countries.

Everyone was treated to the Road Ahead Concert, a whirlwind tour of Indigenous history over the last 100 years. Twenty-five performers, including Yothu Yindi, Paul Kelly, Kev Carmody and Christine Anu, each sang a song.

Dianne Edwards and her kids Murray and Anita take a break from festivities, while Eileen Wingfield and Nina Brown watch over the stall.

Waking up with Dreaming

October

An exhibition featuring the Kungkas' artwork was part of the 2001 Melbourne Fringe Festival. A variety of beautiful handcrafted *punu*—wooden artifacts were showcased in conjunction with screenings of *Waking up with Dreaming*, a film by the Firestick Messengers.

A lot of hard work goes into making the punu—wooden artifacts.

Clapsticks, beads and wira—wooden dish.

Protest threads

November

The campaign broadened with the distribution of colourful Irati Wanti tee shirts, designed by MK Corinne Balaam and printed in Melbourne. The tees soon became a popular fashion item, spotted from Hobart to Broome. And of course Coober Pedy!

Right: Irati Wanti model Cherika Lang.
Over page: Ivy Stewart.

2002

JANUARY 23

INTERNATIONAL DUMP COMPANY LEAVES OZ

Pangea Resources shuts their Australian operation due to widespread community opposition. Company spokesperson, Dr Charles McCombie, says he had not given up hope of "Australians coming to appreciate the high benefits and low risks that would be associated with hosting a well-organised and managed international facility".

Trans/actions

April

The Irati Wanti office hosted Francesca da Rimini as part of the Trans/actions pilot program. Funded by the Australia Council for the Arts, Trans/actions links up new media artists with human rights and social justice organisations in the Asia-Pacific region. Francesca, an Adelaide-based artist, worked with the Irati Wanti coordinators for a month, passing on essential technical skills and industry knowledge to be able to record and promote the Kungkas' artwork.

Francesca da Rimini.

Natalie Austin and Emily Austin with Francesca at the Irati Wanti office.

Trans/actions meeting. Adam Grubb, Kath Hope, Nina Brown, Emily Johnston.

Sammy Brown cooks malu—kangaroo on a bush trip during Francesca's stay.

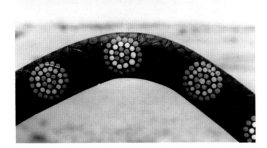

Looking after the culture

The Kungkas are always busy making cultural artefacts.

These are some of the many works documented as part of Trans/actions.

Letter to Mike Rann, recently elected SA Premier (Labor).

May 22, 2002

To Premier Mike Rann,

We write this letter and you can speak for us and we will back you up. We have been talking all the time and no one has been listening. Young people and old people got to say no. Our time is finished, long time finished. Young people have got to talk properly. Why do we have to give them (the Government) the *manta*—earth? *Manta* is in Sydney too. We got everything, bush tucker, *maku*—witchetty grubs, kangaroo meat, honey ant, *mangata*—quandong, wild fruit, wild flower, wild banana. We dig it out and everything, that's why we are hanging on.

Please help us and we will help you. Please, we don't want a waste dump. We are crying for the little ones. Little ones coming up, coming up, coming up. They want to see the old country too. Soft ground, water goes down, down into soft ground. The way they're going on nothing is going to live in this world. Why do they want to bring that waste back? We can't sleep because we are worried for our grandchildren everyday, every morning, every night. What are they going to do? They don't know what's coming to them. Generation to generation we are thinking about the kids, don't you think about your kids? Sit down and look at the kids, crying inside. New baby everyday, another lot of babies. They don't know, they only drink milk and they don't know.

We're old women, we can go anytime but what about the young ones? The sickness is hard enough now. We've seen the cancer. You get bruised up inside and that's what they can't heal. Never mind we tired, we still don't want a waste dump.

We have been talking and talking. Young people got to help us too. Young people stand up and fight strong. Talk properly. Talk strong.

Jamal Witchen, 'Marley Boy'.

We had enough at Maralinga and Emu Junction. They never let people know. Never ask Aboriginal people. We never tell them to go ahead, *wiya*—no. This time we say 'NO'. But they are still coming. We say 'NO'.

We always go hunting, pretty country. They (the Government) are going to spoil it. It's going to get on the back of everything, tobacco, sugarcane. The wind blows it all the time. We are the people living on the land. We're not living in a high storey building. We're living on the flat. We go out camping, make a fire and talk the same about the *irati*—poison, waste dump. We're still talking. We never read the books, but we think about it, right from the heart. We see the little ones running around. We have been talking since a long time ago. Never mind we lost one by one. We are still talking. Can you please help us stop the uranium?

Little trees always grow and when little tree grows they cut them down, it's the same. Spoil them. We are a tree, old tree. We want to see little tree go big, like human being.

Young people got to take it on and they got to be strong. Fight for family, grandchildren, grandchildren's children, and when they grow big they can have children. We look at them crying inside. They may never turn twenty if this spread everywhere. If you can help it some of the kids might not die.

Be strong, don't give them away. You've got to talk properly. You young people think hard, think about your grandchildren when they grow.

We don't want it from the North.

We don't want it from Western Australia.

We don't want it from Sydney.

Keep it in your own place, in your own community. Why they bring it here? Why? This *manta*—ground isn't like Sydney, no *irati*—poison. We have saltwater, dig it out, and drink the water. We drink the water from crab holes, waterholes and dams. Dam finish. Waterhole finish.

You fellas got to listen to us. Please. What that mean? Please? We have been saying please for a long time. We're frightened for the little ones.

We send you this letter, good letter, and you think about us and we be thinking about you too. Write a letter to us please.

From all the Ladies,

Kupa Piti Kungka Tjuta

Tiffany Brown, Georgia Brown, Mia Brown.

"We're crying for the little ones. Little ones coming up. They want to see the old country too"

Postcard to Coober Pedy

July

The Kungkas released 5,000 Avant Card postcards reiterating their opposition to a radioactive waste dump in their *ngura*—country. The national postcard campaign coincided with the Federal Government's release of the draft Environmental Impact Statement (EIS).

The postcard became a vital campaign tool, distributed nationally and internationally. Rather than sending them to the Government in protest, the Kungkas requested they come home to Coober Pedy. People from all over the world sent messages of support to the Kungkas.

Later in the year a further 90,000 cards were printed for distribution.

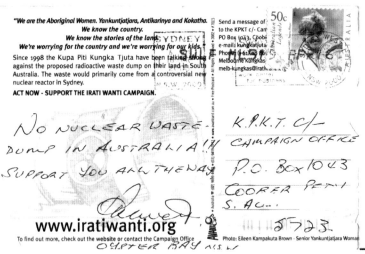

JULY 27

DRAFT EIS RELEASED

Federal Government releases the draft Environmental Impact Statement.

At this stage Site 52a is still the preferred site, but nearby sites 40a and 45a in the Billa Kalina region still remain under consideration. The draft EIS calls for public submissions about the dump plan, before a final EIS is prepared.

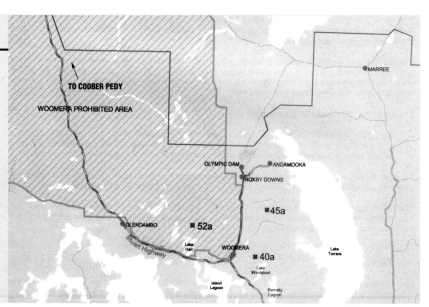

AUGUST 16

SA GOVERNMENT BANS DUMP

Recently elected SA Labor Government responds to substantial community pressure and legislates against any waste dump being built in the state.

"It's our nuclear deterrent. If any future Federal Government moves to override the state laws banning a nuclear waste dump by using its constitutional powers, that will immediately trigger a referendum of all South Australians." Mike Rann, SA Premier (Labor).

Trade unions on board

September

At the Australian Council of Trade Unions (ACTU) Indigenous Conference in Brisbane delegates unanimously passed a motion, moved by John Hartley, acknowledging "the absolute right of the Kungka Tjuta - Elders and custodians, to assert their lawful protection of country". The conference appealed to the ACTU to mount a coordinated campaign to oppose the waste dump.

The South Australian branch of the United Trades and Labour Council (SAUTLC) answered the call in March 2003 with 30 affiliated unions unanimously voting to ban construction and the provision of services to any waste dump project in SA. It added weight to a similar ban that had been recently imposed by the Construction, Forestry, Mining and Energy Union (CFMEU).

Later, in August 2003, a 'priority policy' was passed at the ACTU National Congress. "The ACTU and affiliated unions stand in solidarity with the Indigenous communities in opposition to the nuclear waste dump."

Top: Karen Crombie, Eileen Crombie, Eileen Brown, Sandra Taylor, Emily Austin, Dianne Edwards. Bottom: Maggie Ward, Cissie Riley, Myra Watson, Angelina Wonga, Martha Edwards. Banner painted by Lyn Hovey.

Secret Country

September

An art exhibition in support of the Kungkas opened at the Gould Gallery in Toorak, Melbourne. In July, six prominent Australian artists and a photographer had ventured into the Woomera Prohibited Area with Kokatha man Andrew Starkey, to visit the proposed dump site. After meeting the Kungkas in Coober Pedy and hearing their stories, the artists united in their desire to assist the Kungkas' fight. *Secret Country* publicised the campaign to a broader audience and raised much needed funds.

MK Camilla Pandolfini talks at the opening.

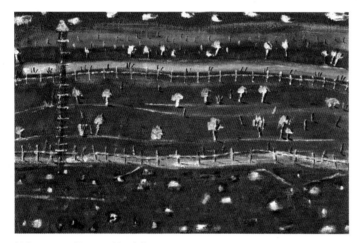

Woomera Tanya Hoddinott.

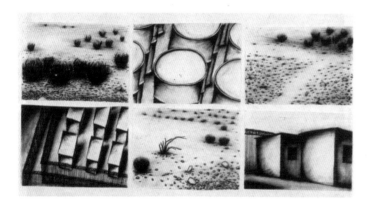

Woomera 1-6 Greg Ades.

Rock legend Paul Kelly, anti-nuke legend Dave Sweeney.

2003

JANUARY 23

FINAL EIS RELEASED

"It is a fact that the majority of the [667] submissions [to the Environmental Impact Statement process] were opposed to the siting but a lot of them were on erroneous grounds." Peter McGauran, Federal Science Minister.

The Final EIS goes to David Kemp, Minister for the Environment, for approval.

Order of Australia

January

On January 26, Australia Day, Eileen Kampakuta Brown was awarded a Member of the Order of Australia.

Mrs Brown's award was for service to the community "through the preservation, revival and teaching of traditional *Anangu* culture and as an advocate for Indigenous communities in Central Australia".

Eileen Kampakuta Brown.

JANUARY 31

BIG BUDGET TO SELL DUMP

Multinational public relations company, Hill and Knowlton, win a $300,000 Federal Government contract to market the dump plan to the SA community.

"The Government will constantly tailor the campaign so as to provide the public in SA with all the facts to dispel the constant misinformation and distortions being circulated." Peter McGauran, Federal Science Minister.

FEBRUARY 3

CONTROVERSY OVER SITE 52A

The Australian newspaper reveals that Defence Department officials are "violently" opposed to construction of the dump at Site 52a, which borders on a military weapons target range. The Department is highly critical, saying "the EIS was misleading, failed to adequately consult Defence and misjudged missile impact risks and radiation exposures".

Right: Crashed Jindevuk Spy Plane near Site 52a.

Honoured, ignored in the same day

Koori Mail
February 12, 2003

Not many people get honoured and ignored all on the same day but that is what happened to senior Yankunytjatjara/Antikarinya woman Eileen Kampakuta Brown, a member of the Kupa Piti Kungka Tjuta from Coober Pedy, South Australia.

On January 26, Mrs Brown was honoured as a Member of the Order of Australia just days after the final Environment Impact Statement for a radioactive waste dump in South Australia was released. The waste dump is a project that Mrs Brown has opposed for five years.

Mrs Brown's award is for service to the community 'through the preservation, revival and teaching of traditional A<u>n</u>angu [Aboriginal] culture and as an advocate for Indigenous communities in Central Australia'. Her 'curriculum vitae' shows a sustained commitment to the protection and restoration of sacred sites; involvement in native title claims; and the protection of women's and children's rights.

Mrs Brown has been recognised as a senior woman of extensive traditional cultural knowledge. And she says it is this very cultural knowledge that has compelled her to lead the struggle against the Federal Government's proposal to dump radioactive waste in the South Australian desert.

The release of the final Environment Impact Statement was accompanied by a Federal Government announcement of a $300,000 're-education' budget to address the 'concerns' of the South Australian public. The dump faces strong opposition in South Australia.

Mrs Brown says she will continue to speak out against the dump. "I've got the knowledge. Never mind that I don't speak English, I speak strong," she said. "They [the Federal Government] don't listen. They got no ears."

Angelina Wonga cuts punu—wood.

We of Little Voice

March

We of Little Voice, a documentary featuring some of the Kungkas, premiered at the Adelaide International Film Festival.

Directed by Arabunna man Peter Hodgson, the film follows anti-nuclear campaigner and Arabunna Elder, Kevin Buzzacott, as he travels over gravel roads listening to Elders who have experienced the effects of the nuclear industry.

The documentary screened on SBS TV in August.

Film screening at Umoona Aged Care, Coober Pedy.

APRIL 14

ENVIRONMENT MINISTER IGNORES OWN ADVISORS

The Department of Environment's Indigenous Advisory Committee writes to David Kemp stating, "The Kupa Piti Kungka Tjuta fundamentally oppose this nuclear waste dump which they see as the imposition of poison ground onto their traditional lands. The Kokatha People, as registered Native Title claimants, oppose the nuclear waste dump and the intended acquisition and annulment of their native title rights and interests."

"Throughout the EIS process... the Native Title claimants and other community members feel that there has not been adequate consultation. Traditional Owners have also not been able to find out about the intended legal approach of the Commonwealth Government in carrying out key aspects of the proposed project."

Off to Adelaide
April

Launch of *My Young Life*

An evening of film screenings and a book launch, sponsored by the United Trades and Labour Council of SA, was held at the Tandanya Aboriginal Cultural Centre. Eileen Brown and Sister Michele Madigan made the *tjapu*—small book. *My Young Life, An Anangu—Aboriginal Love Story* is about Mrs Brown's childhood, walking the country and meeting her husband at Mt Willoughby Station.

Ngitji Ngitji Mona Tur interpreted throughout the book launch.

Dr Irene Watson reads an extract from the tjapu— small book.

'Every morning, *unkara tjuta*—we young girls used to swim in the waterhole. This day, I had a petticoat on and a pretty dress. There was somebody sitting down near. Eileen Crombie and I, we were swimming in the waterhole. I was the last one to go in. And afterwards, while we were swimming, we saw who it was sitting there; it was Tommy Brown.' *My Young Life*, pg 17.

Government House

The next day Mrs Eileen Brown was awarded her Member of the Order of Australia at Government House by the Governor of South Australia, Her Excellency Marjorie Jackson-Nelson.

Eileen Brown receives her prestigious medal.

Adelaide continued...

Parliament House

South Australian women politicians hosted an afternoon tea for the Kungkas in honour of their many achievements. In a tearoom at Parliament House women from Labour, Liberal and Democrat parties put their political differences aside to enjoy the special occasion.

Afternoon tea with politicians.

Labor Senator Penny Wong with Eileen Brown and Emily Austin.

Following afternoon tea the Kungkas were greeted by a large contingent of media and supporters.

The Kungkas with Labor MP Lyn Bruer.

Back: Eve Vincent, Emily Johnston, Nina Brown, Simone Tur, Martha Edwards. Front: Emily Austin, Mona Tur, Eileen Brown, Eileen Crombie.

On the road again

After the events in Adelaide, Eileen Brown, Emily Austin, Eileen Crombie, Martha Edwards, as well as Nina Brown, Emily Johnston and MK Eve Vincent travelled 2600 kms across western NSW to Sydney. The seven women drove from Adelaide, along the Barrier Highway through the Blue Mountains, the same route the radioactive waste would be transported if the dump were built in SA.

Kungkas win global greenie award

From San Francisco...

Eileen Wingfield with Richard Goldman and other prize winners at the official award ceremony.

April 15

Eileen Wingfield and Eileen Brown were awarded the prestigious American-based Goldman Award for the Environment. The Prize is given annually to "grassroots environmental heroes" from six geographic areas and is considered the greenie Nobel Prize. Both women were invited to the USA to collect their prize.

Eileen Wingfield and her daughter Rebecca Bear-Wingfield flew over to San Francisco for the official ceremony and met with the five other prize winners from around the world. The group then participated in a whirlwind 10 day tour, attending a round of news conferences, media briefings and high-level meetings in San Francisco and Washington DC.

This was Mrs Wingfield's first trip overseas. Despite ill health and extensive family and cultural responsibilities she undertook the massive journey, attesting to her courage and resolve to protect Kokatha country for future generations.

On a full page in the *New York Times* Mrs Brown and Mrs Wingfield were honoured with the other winners for their exceptional courage and commitment to protect the environment.

Eileen Wingfield and Rebecca Bear-Wingfield at the Golden Gate Bridge.

To Sydney

Coinciding with the award ceremony in San Francisco, Eileen Brown was the guest of honour at a special ceremony at the Sydney Observatory overlooking the harbour. The ceremony was attended by local Aboriginal Elders, media, politicians, family, friends and supporters of the Kungkas, including Christine Anu who presented Eileen Brown with her award.

Karina Lester, Eileen Brown's granddaughter, flew over to Sydney from Adelaide to interpret for the Kungkas.

Eileen Brown and Christine Anu.

While driving to Sydney the Kungkas composed a special Inma about protecting the country from the irati—poison for future generations. They sing it for everyone at the award ceremony. Karina Lester, Eileen Brown, Emily Austin, Eileen Crombie, Martha Edwards.

Greens Senator Bob Brown, former Goldman Prize winner, with Eileen Brown and Karina Lester.

Journalists interview Eileen Brown after the ceremony.

MKs receive a Golden Egg to acknowledge their contribution to the campaign. Eve Vincent, Shannon Owen, Clare Land, Breony Carbines, Sam Sowerwine, Corinne Balaam.

Kungkas with Wiradjuri Elders Sylvia Scott and Isobelle Coe.

Sutherland Shire visit

The Kungkas met with Sutherland Shire councillors and community members who are active campaigners against the construction of the new nuclear reactor at Lucas Heights.

Karina Lester addresses the gathering at the council chambers.

Phil Blight, Sutherland Shire Mayor, with the Kungkas.

Letter to David Kemp, Minister for the Environment, only days before he was expected to approve the dump EIS.

May 7, 2003

To David Kemp,

We are the Kupa Piti Kungka Tjuta - the Senior Aboriginal Women of Coober Pedy. We heard you are making up your mind about the dump. We have written many letters over the years talking over and over about this nuclear dump business and you still don't listen.

We have just come back from a big trip to Sydney because two of our Ladies won the big Goldman Prize for all the work we have been doing to stop the dump and save our culture. We went in the car across the highway, the same one those trucks carrying the *irati*—poison would come. We know that road now and it's too dangerous with all those people in Broken Hill, Dubbo and the Blue Mountains.

You have been playing around with that poison for a long time in Sydney at Lucas Heights. You have to keep that rubbish there where you make it. You can keep an eye on it there, but here you would just forget about it and it will poison the water.

We are the Culture Women and we are worried about the *tjitji tjapu*—all the little kids still coming up after us. But we are worried about everyone's kids, not just our own. The beautiful desert is our home.

Angelina Wonga, Tjunmutja Myra Watson, Eileen Kampakuta Brown, Emily Munyungka Austin.
(pictured below)

Kupa Piti Kungka Tjuta

"We know that road now and it's too dangerous"

MAY 9

DUMP EIS APPROVED

Amid controversy due to its proximity to the live rocket range, Site 52a in the Woomera Prohibited Area is abandoned. David Kemp approves the Final EIS for Site 40a on nearby pastoral property, Arcoona Station. "My decision follows a rigorous and transparent assessment process with full public involvement."

As the Government explained in the EIS, it will now move to acquire the dumpsite by using the Commonwealth *Lands Acquisition Act 1989*.

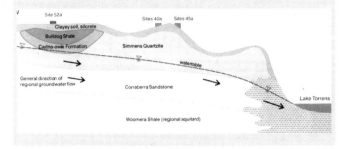

Eileen Brown digs for maku—witchetty grub with her granddaughter Karina Lester.

Living Black

May

Eileen Crombie, Emily Austin, Eileen Brown, Myra Watson, Angelina Wonga and family members appeared on SBS current affairs program *Living Black*, in a feature produced by Tanya Denning.

The film crew arrived in Coober Pedy the day after the Federal Government announced the approval of Site 40a on Arcoona Station. Everyone headed to Cadney Park, 150 kms north of Coober Pedy, and spent the day collecting *maku*—witchetty grub with grandkids, and filming interviews against Cadney's beautiful red sand backdrop. Karina Lester interpreted for the Kungkas.

Yankunytjatjara Elder Yami Lester and Kokatha man Andrew Starkey also featured in the program.

Dimos Tsakaridis shows producer Tanya Denning how to get maku out of the plant root.

Tiffany Brown with maku.

Letter to Mike Rann, SA Premier (Labor).

May 26, 2003

Dear Mr Rann,

We all fight together.

We don't want the poison to come back this way, Arcoona Station. We're still talking strongly about water, we've got spring water. That's all the water we've got underground.

They've got to keep it, the poison, in Sydney. We never told them to bring it back here when they were finished with it.

When I saw you in the news talking against that poison now, the radioactive dump, I was really happy. You was talking strongly. Don't let it come this way. A truck might have an accident. Small towns are halfway, all the way to Port Augusta.

You help us, fighting against it, and we'll back you up. We had enough at Maralinga, they been sneaking that time, they never let us know about that Maralinga bomb.

Keep fighting! Don't give up and we won't give up. Keep fighting because kids want to grow up and see the country when we leave them, when we pass on. They'll take it on. Hope they'll fight like we fellas for the country. We don't want to see the *irati*—poison come back this way. We're not going to give up.

The other lot, Canberra Government, if they're for us they should stick up for us and they're giving us away! Like when the dogs, dingoes, hang around the paddock, they throw the poison. Now today we're waking up, we don't want poison to us. Must be good fun for them, but we don't want it.

Thank you for helping us, but you're helping your kids too, my kids and your kids. We're fighting for all the kids. Thank you very much. I might see you again on the TV.

Eileen Unkari Crombie

One of the Kupa Piti Kungka Tjuta who always sticks up for a man like you 'til the end of the world

"Keep fighting! Don't give up and we won't give up."

Eileen Crombie with her great grandson Adam Buzzacott.

JUNE 2

LEGAL LOOPHOLE TO STOP DUMP

SA Premier Mike Rann declares he will use Section 42 of the *Lands Acquisition Act 1989 (Cth)*, which says the Commonwealth cannot acquire a public park without the state's consent. Rann drafts legislation to declare Site 40a on Arcoona Station a public park.

All part of the job. Emily Austin shows Nina how to hunt ungkata—frill neck lizard, then coaches her through the fine skills of cooking them up the right way.

Radioactive recognition

June

On June 5, World Environment Day, the Australian Conservation Foundation (ACF) announced the winner of the 2003 ACF Peter Rawlinson Conservation Award: Nina Brown.

The award acknowledges the outstanding voluntary contributions of individuals and groups to the conservation of the environment.

"This award recognises Nina's commitment, tenacity and importance in helping to ensure the community voice is heard," said ACF executive director, Don Henry.

JULY 7

URGENT LAND GRAB BY FEDERAL GOVERNMENT

The very day the SA Government tables the Public Parks Bill for discussion, the Federal Government announces the compulsory land acquisition of a 6.5sq km parcel of land, including Site 40a, on Arcoona Station. Nick Minchin, now Federal Finance Minister, uses the urgency clause, which cancels out all native title, state and pastoral rights. "Our Government could not allow a cynical Mr Rann to prevent the safe and responsible management of Australia's radioactive waste."

Minchin maintains dump construction will begin within 12 months. "We now have the land and once the licence is obtained... the repository [dump] can be constructed."

Just days after the land acquisition Nina Brown and MK Breony Carbines join hundreds of greenies in Adelaide and march to Senator Nick Minchin's office in support of the Kungkas.

Letter to Peter McGauran and Nick Minchin following the Federal Government's compulsory land acquisition of Site 40a.

July 8, 2003

To Peter McGauran and Nick Minchin,

You don't listen to us ladies. You're still not listening. Do we have to talk over and over?

He (John Howard) should have come and faced us, have a meeting, talking and things.

John Howard jumping around all over the place, over the world. He should be at home looking after us. One Government, one man, and he doesn't listen to us. He goes to war, kills people, he's doing that here too. We know what's going on. We've been around.

Listen to us. We women have the rights of this land. We protect it. We will write a letter to the Queen to help us. Tell her to come over to Australia. Howard won't listen to us. Our Government don't listen. They bought that ground.

Government say 'fair and just compensation'. We don't want money. We weren't born with money. We want life, land, for the kids. They just put that word 'money' in. Trying to buy us. We want the life. Kids' life, our life. That's white man's money.

It's women's place. Stop mucking around with women's business. It's our story to know for all *kungkas*. Not a story for you white men. Not your land even if you say you own it. Even if you buy it. It's women's place. It's Dreamtime from long time history. We keep the story. The land holds the story, not you, spirits are still there. Stop mucking around with women's business.

You're digging a hole in the Dreamtime. If you dig this hole in the *manta*—earth and fill it with the poison, make the dump, something will happen. There will be anger. If you don't listen you will be sorry. We talking and talking, go round and round same words. We're trying to help everyone. We talking straight. Don't go there. It's dangerous.

In the beginning Aborigines were given the land. Not white people. You just gotta *kulini*—listen. We aren't just some grass under a rock.

*"Not your land
even if you say you own it.
Even if you buy it.
It's women's place."*

Where do you live? Not in space. You live in Australia too. On the *manta*—earth. Put it in Canberra with you. You look after it.

Listen, look out after us. You've got sons and daughters too. We're crying out for help. Please listen. Don't poison us. We are pleading to you. You've got families, same as us. We need to protect them all. So do you. We're not being cheeky with you. Please help us. We're not looking for a fight. We shouldn't have to fight for our land, just to get rid of the poison.

Please no poison. We got water and bush tucker; kangaroo, emu, bullocks. What about the bullocks and the sheep? That's farming country too, they come from the station. What will happen when they are poisoned? Emu drink same water. Kangaroo, goanna, perentie, cattle and sheep, all drink the same water. Then we eat them, like you. The water will poison the animals and kill them all, then you fellas and us.

No more. Just leave it now. We've got a lot of spirit in the land. You have taken so much from us. Leave us now. We tired now. We want to live in peace. Please listen this time. Time is getting shorter. The Lord put us here to look after the *manta*—ground. We're greenie *mula*—true greenies. We're here to look after the whole earth.

Ivy Makinti Stewart, Eileen Kampakuta Brown, Eileen Unkari Crombie, Emily Munyungka Austin, Tjunmutja Myra Watson, Dianne Edwards.

Kupa Piti Kungka Tjuta

"We bring our worry" to Silverton

July

Barkenji, Adnyamathnya, Arabunna, Yankunytjatjara and Pitjantjatjara Elders, all of whom have been affected by the nuclear industry, came together with greenies at the No More Radioactive Genocide gathering in Silverton, outback NSW.

Eileen Brown, Eileen Crombie, Emily Austin, Myra Watson, Jeannie DeRose and Karen Crombie, were accompanied by Clare Brown, Georgina Wright and MK Breony Carbines.

The recent land acquisition fuelled a sense of urgency and the Kungkas called on the gathering:

"We Kungka Tjuta were sitting down doing our own things, our work in Coober Pedy, and we were invited to come down here to Silverton."

"We bring our worry, the waste dump down here for support."

"We're frightened for the transport of the waste through the Blue Mountains, Dubbo and Broken Hill. If the truck has an accident it will kill everything. The kids on the road, the animals, everything. We're giving you fellas an idea, the knowledge, listen. We bring it from the heart. We want people to block the road. Don't let anything through, not the trucks, block the trucks and the trains, check them."

"Leave the waste in Sydney. Keep it!"

"We got to fight together and we'll be strong. We can work together. You help us and we'll help you."

"We got to fight together and we'll be strong." Myra Watson and Eileen Brown.

Adnyamathnya Elder Ron Coulthard, who has actively fought against the Beverly uranium mine on his country, said, "I reckon we should forget about Beverly for a couple of months. I think we will go on with this waste dump... We can all keep our eyes on the public roads... And like the Old Ladies said, we got to help each other. And it doesn't matter where we come from, brown, yellow, black or white, let's get together and pull together."

Atsuko Nogoua, a Japanese woman who works in support of survivors of the Nagasaki and Hiroshima bombs, presented each Kungka with paper cranes in recognition of Japan and Australia's shared bomb histories.

During the gathering the Kungkas called all the women to sit down to discuss women's important responsibilities to country.

Arabunna Elder Kevin Buzzacott and Adnyamathnya Elder Ron Coulthard.

In recognition of shared bomb histories, Atsuko presents Emily Austin with peace cranes.

Women raise their hands in support.

"Sit down all together and listen to the story"

During their time at Silverton, the Kungkas decided it was time to have a big meeting in Coober Pedy.

After returning home they offered an open invitation for everyone to come and sit down and "talk about the poison" at Ten Mile Creek Elders Camp.

On the way home. George Wright, Clare Brown, Eileen Brown, Eileen Crombie, Emily Austin, Karen Crombie, Jeannie DeRose, Breony Carbines, Myra Watson.

"I can assure you" says Kemp

July

David Kemp, Federal Environment Minister, wrote to the Kungkas in reply to their letter dated May 7, 2003, regarding his approval of the final EIS for the proposed waste dump. It was the first correspondence received from a Federal Government Minister since the Kungkas launched their campaign in 1998. Below are extracts from the letter.

"I fully understand your concerns about the possible 'poisoning' of lands near the repository site, particularly when taken in the context of the British atomic tests at Maralinga and Emu. However, I can assure you that the low-level waste repository is in no way comparable to the atomic tests, and that the facility will be well-managed and controlled."

"I can assure you that the government has listened to your concerns about the proposed repository."

"I can assure you that only low-level waste such as from material used in hospitals will be stored in the new repository, not waste from Lucas Heights."

"I can assure you that the government will not forget about the repository and there will be no poisoning of groundwater from operation of the repository."

Dr David Kemp, Federal Environment Minister

July 16, 2003

Hundreds of posters and invitations were distributed to the Kungkas' families, A<u>n</u>angu communities and wider campaign networks.

Japanese professor moved by shared history

September

During hectic Bush Camp preparations the Kungkas took time out to host an important international guest.

Professor Ryoichi Terada came across the Irati Wanti website as part of his research into international nuclear issues. He travelled from Japan to Australia especially to meet with the Kungkas.

Ryoichi told the Kungkas that he was overwhelmed by the desert while driving the 850 kms from Adelaide to Coober Pedy alone in a rental car. It was a stark contrast to his five hour return train journey to and from his university, through suburban Tokyo everyday.

Ryoichi and the Kungkas spent an afternoon sitting around a campfire sharing their knowledge of atomic bombs.

Myra Watson and Professor Terada exchange gifts after a powerful meeting.

The making of Bush Camp

September

With the invitation now circulating around Australia and weeks to go before the convergence at Ten Mile, a sea of handwritten lists threatened to overtake the office. The task ahead was huge and exciting: fundraising, organising transport for *Anangu* from across SA, making a website and book, ordering food and finding catering crews, and ensuring Ten Mile would be a safe and functional campsite for hundreds of people.

76   *Talking Straight Out*

A week before Bush Camp began kids from Coober Pedy Area School's Aboriginal Education Class spent the afternoon at the Irati Wanti office designing and painting the signs to be placed around town to direct all the visitors to Ten Mile.

Sign artists: Tiffany Brown, Denise Lennon, Lucille Warren, Kaylene O'Toole, Anna Tsakaridis, Trudy Brown, Vincent Warren, Norman Riessen, Jeremy Williams. And teachers: Pearl Austin, Pat Clifford, Barry Daniels, Elaine O'Toole.

'Kulini Kulini' Are you listening?
Bush Camp at Ten Mile Creek September 29 – October 1

Bush Camp began with the sad news of a local *Tjilpi*—male Elder passing away. Rather than cancel, the Kungkas decided to move Sorry Business to Ten Mile. The news itself was testament to the constant presence of loss and grief in the Kungkas' lives, and their choice to press on with the camp showed their resilience and commitment to looking after the country.

Having travelled far and wide to spread their Irati Wanti message, it was time now for everyone to leave the comfort of their own homes and come to the Kungkas' place. *Anangu* came from Ceduna, Port Lincoln, Yalata, Oak Valley, Port Augusta, Mimili and Indulkana to join the Kungkas.

By Monday over 300 people had passed by the registration caravan. They arrived by car, bus, train, plane and bike. Over the next three days, families, friends, supporters, community groups, locals, greenies, media, and politicians all gathered under a marquee amidst swirling dust, gale force winds and storms, to sit on the *manta*—ground and *kulini*—listen.

Elders recalled their experiences of the atomic bomb tests in the 1950s and 60s and their impact on their families, health and lives. Half a century later, the waste dump posed a similar threat to people's lives and land. The courage and generosity of the Elders who spoke out at Bush Camp inspired a collective will to support the Kungkas' campaign to stop the *irati*—poison from returning.

"I wish John Howard was here to Kulini Kulini... He was scared of the old Kungka Tjutas. He just don't like to listen. He know it's best for him, not for the Kungkas, for himself. I know he made a lot of money in that poison but he's going over the mark, he's getting too greedy."

Emily Austin, Coober Pedy

"Fifty years is a long time for sadness. In fifty years they have kept this quieter than anything, and nobody never told us they were going to drop the bomb. We must stand strong and fight."

Myra Watson, Coober Pedy

Tjilpi Tjutaku Karu—All the Old People at the Creek

Ten Mile Creek Elders Camp

In 1997 Umoona Aboriginal Aged Care Corporation received funding to set up a traditional bush camp outside of Coober Pedy at Ten Mile Creek.

The Creek is an ancient songline that has been travelled by foot for generations by the Old People from the Anangu Pitjantjatjara Yankunytjatjara (APY) Lands to Oodnadatta. Along the way, families would camp at the site which is a natural windbreak area and water was collected from a soak nearby. It is widely accepted as 'everybody's land'.

Source: *Bush Camp Program*, Umoona Aged Care 1998.

"My name is Eileen Kampakuta Brown. I'm still saying no. I'm a grandmother. I've been sick but I'm better and I'm still talking, I'm still strong. I've lost my sister, my cousin, but I'm still talking strong."

Eileen Brown, Coober Pedy

"I been fighting Roxby [uranium mine] since the start. We don't want the dump because we've seen what happen to our children [from the bomb]. Everyone was sick. I got a grandchild, he's got a tumour in his head. Why they do this to the innocent?"

Eileen Wingfield, Port Augusta

"We were sitting down near Maralinga, in Yalata, when they drop the bomb... We didn't know what that was, some sort of an earthquake or something like that. When we got up and heard one men said, 'something is happening', and we never knew what that was. Then we got sick."

Alice Cox, Oak Valley
(pictured on right with Angelina Wonga)

"I was working in Maralinga. I loaded up blankets. After working I got crook everywhere, asthma, got really sick. Government never told me about Maralinga, they never. I had another wife... My wife finished, she died. I was left alone."

Larry Crombie, Oak Valley

"I come from Batchelor. I'm from the Kunarakan tribe. That's our traditional land there and it's right where the Rum Jungle [uranium] mine is.

In the end, the poison and that, the contamination, it's not prejudice. It's going to take us all, black and white. When people are talking about reconciliation between black and white, this is the place that it should be happening. Here, out here, where they want to put the poison. And there should be thousands of people here."

Speedy McGuiness, Batchelor

Karina Lester interprets for many of the Elders during the meetings. Mutintja Prince, Willy Tinyku, Jimmy Bannington, Billy Mungie, Karina Lester.

Journalists from major newspapers, radio outlets, and independent film crews ensure the stories told at Bush Camp are heard by a wider audience.

Painting in the kids shed with Natalie Austin, Emily Austin's daughter.

Greenies split into regional groups to discuss how they could best support the campaign.

Singalong. Ian Crombie, Bobby Brown, Eileen Crombie, Emily Austin, Eileen Brown.

Maisie Wintinna, Eileen Crombie's daughter-in-law reads an extract from the book.

Launch of *He Was a South Australian Film Star*

On Wednesday morning everyone attended the launch of Eileen Crombie's first book, *He Was a South Australian Film Star. My Life with Billy Pepper – Tinyma.* Sister Michele Madigan worked with Eileen Crombie to record the small memoir of her husband's life. Eileen's sons, Ian Crombie and Bobby Brown, played music and sang songs. It was a great success and many books were bought and signed.

> 'Billy Pepper was working in Twins Station and from there he met me then in Coober Pedy. I was single; we were only boyfriend and girlfriend then and I left him then; we were going to Yalata, being moved from Bulgunnia Station - moved for the Maralinga bomb. The policeman came to Bulgunnia Station and took all the *Anangu*—Aboriginal people, and moved a whole truckload of us to take us to Koonibba Mission, other side of Ceduna.' *He was a South Australian Film Star*, pg 4.

Twelve women identified by black and red arm bands were GANG (Girls Against Nuclear Genocide). From Elders support and media liaison to first aid and rubbish runs, GANG was responsible for camp logistics.

Emily Johnston, Karrina Nolan, Shae Clayton-Freedman, Clare Brown, Suzanne Woolford, Alex Kelly, Nina Brown, George Wright, Sam Sowerwine, Lou Bolt, Meg Jones, Breony Carbines.

GANG meetings were held twice a day to keep everything running smoothly.

Wood collection to keep the waru—fire burning and everyone warm.

Umoona Aboriginal Aged Care cook, Mary Harms, coordinated the *kuka*—meat kitchen and worked non-stop over three days to prepare hearty meals for *Anangu*.

Mary Harms directs the greenie boys where to place the Hangi, a traditional Maori earth oven. Meat sourced from Mabel Creek and Mt Willoughby, nearby Aboriginal-owned stations, was cooked in the Hangi.

FoE Melbourne coordinated the *mai putja*—non-meat food kitchen, supplying the greenies with wholesome tucker.

Inside the mai putja shed.

Mai putja cooking collective.

Karina Lester and Angelina Wonga.

To close Bush Camp the Kungkas invited everyone to sit down for *Inma*

Karina Lester, Eileen Brown's granddaughter explained *Inma* to the crowd.

"*Inma* is so important in passing on the song and dance for this area and if you start playing around with land, people will lose their songs and dances. That's why these old Kungkas here are singing, and singing strongly. Doing the dances because they know the Seven Sisters stories."

"They know singing for the land and doing the dance for the land is healing the land. We work together, partnership between *Anangu* and land and if we look after land through our songs and our dances, that's the way we can continue on."

"If things are changed and taken out of our land our stories are broken down. The learning will not continue and that is why Nanna Mob are speaking strongly about it, singing these songs, doing these dances on this land so it still stays strong. That's something that needs to be passed on to the next generations to come because it is so important to us as *Anangu* people."

Inma opens with Willy Tinyku, a Yankunytjatjara Elder, telling the story of Wati Nyiru—lover boy, who followed the Seven Sisters all over the country.

Angelina Wonga.

Martha Edwards.

After the opening the Kungkas called up all the women to learn the dances

Letter written on the eve of the 50 year anniversary of the bomb - the first British atomic bomb tested on the Australian mainland.

October 14, 2003

"Enough is Enough"

We are writing from Coober Pedy, South Australia. We are the Old Aboriginal Ladies.

When the smoke from the bomb came over, we were young girls. We were walking around the country as little girls with our families, living off the land, working on stations. We were happy, no worries. The country was clean. No sickness, no asthma.

When that bomb went off it shook the ground and the smoke, the poison, came over us. The poison spread through the country. All the people got sick and everyone is still sick. Sore throats, sore eyes, asthma, blindness and girls can't have children. We can't sleep, too much pain.

We all say enough is enough. *Irati wanti*—just leave the poison. Please listen. We're tired. NO NUCLEAR DUMP. We have had enough. We're stressing out with too much worry. We are always talking about the *irati* and alongside we are always having bad luck with people passing away. Stop the *irati*, we can't take it anymore.

We are talking for everyone and we are fighting for our grandchildren and our great grandchildren's future now. We'll be gone.

Eileen Kampakuta Brown, Emily Munyungka Austin, Tjunmutja Myra Watson, Martha Uganbari Edwards, Eileen Unkari Crombie, Angelina Wonga, Maggie Ward, Lily Baker.

Kupa Piti Kungka Tjuta

Eileen Crombie with her great granddaughter Roselle Crombie, and Minnie Toby.

"We were all living when the Government used the country for the bomb"
Remembering 50 years

On the morning of October 15, 1953 the British military detonated an atomic bomb called Totem One at Emu Junction, a flat claypan 280 kms north-west of Coober Pedy.

This test was the first nuclear test conducted on the Australian mainland as part of a British weapons testing program, which ran from 1952-1963, and received the full support of the Australian Government and involvement of Australian Army personnel.

On the morning of October 15, 1953 Totem One's cloud rose in the sky. Because of unusual weather conditions it drifted north, without its radioactive particles dispersing. The dark cloud was still visible 24 hours later.

Yankunytjatjara oral histories tell of a deadly black mist, produced by Totem One, which came rolling through the scrub. The black mist devastated Yankunytjatjara camps, causing sudden deaths as well as immediate and long-term sickness.

Angelina Wonga, who was camped at Wanytjapila remembers, "I was with my mother, father and sister, all the family. Sitting down. And when we seen a bomb went out from the south. And said, 'Eh, what's that?' And when we see the wind blowing it to where we were sitting down. Nobody got a warning, nobody. That was the finish of mother and father. They all passed away through that. I was only there. Buried the grandmother. I was the only one left. I lost everything."

"Nobody got a warning. Nobody." Angelina Wonga.

"Right here the smoke caught us, it came over us."
Eileen Kampakuta Brown.

Eileen Brown remembers the morning after the test, "We got up from the tent... everyone had red eyes. Everyone had red eyes. Right here the smoke caught us, it came over us. Us lot... We tried to open our eyes in the morning but we couldn't open them. [We had] red eyes and tongues and our coughing was getting worse. We were wondering what sort of sickness we had."

In 1984-5 a Royal Commission was held into the tests. It found that the decision to detonate on October 15, despite the weather conditions, knowingly exposed *Anangu* communities to extremely dangerous fallout. But the Royal Commission only gave compensation to a few individuals.

After two tests at Emu Junction, the tests moved further south to the Maralinga Range. Southern Pitjantjatjara people were forcibly removed from their land in the testing area, and moved to Yalata on the coast. Myra Watson was born in this area, but was away in the Riverland when the bombs went off. She recalls, "In 1970 I went back looking for my people, they were in Yalata. They sent them to Yalata. I went to Yalata to find them. But there were only a handful there. There used to be big tribes of people. Camps used to be all over, you know. I only see the handful there in Yalata. They put the bomb there. Right through and finished all our people, in the Victorian desert. You look at it on the map, nobody living in the Victorian desert. All our people gone."

Throughout the Irati Wanti campaign the Kungkas linked their memories of atomic tests with the proposed radioactive waste dump: linking past and present. The Kungkas knew from their experience that the effects of the *irati*—poison, are long lasting. As Eileen Crombie said, "The Government might think it is safe, but here in South Australia... we've already lost a lot, through the bomb. That's the one they were sneaking through. This time we've got to stand up and fight for every right."

In memory

On October 15 events across Australia recognised the bomb history, and the Kungkas ongoing fight against the *irati*—poison. Supporters of the Kungkas organised commemorations in Alice Springs, the Blue Mountains, Brisbane, Melbourne and Adelaide.

Radio segments compiled at Kulini Kulini Bush Camp were played throughout the day on Triple J and Radio National.

In Coober Pedy a barbecue was held at Ten Mile Creek.

Genocide Corner, Adelaide.

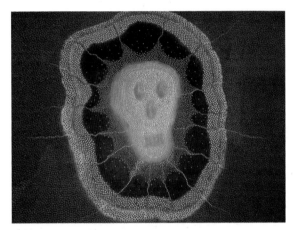

Maralinga, Natalie Austin.

Totem 1 art project

On October 15 an exhibition opened with art on display in shop front windows, display cases and poster windows throughout Melbourne's CBD. Featuring diverse works from 15 local and interstate artists, the collection hoped to generate broader recognition of the painful legacy of the bomb.

Over 60 people attended the opening, walking through the city carrying small silent bells on black ribbon; symbolising the lack of warning and protection given to *Anangu* and Australian Army personnel.

The exhibition was part of a long-term community art project. From 2001-2003 MK Breony Carbines organised workshops at community events where people created art in response to the Kungkas' bomb testimonies. Work from Natalie Austin, Emily Austin's daughter, and Michelle Newman, daughter of an ex-serviceman, were also on display.

Artwork displayed in shop front window.

Sketch by Thea Webb.

Senate motion

The Australian Senate passed a resolution on October 15. Penny Wong, SA Labor Senator, presented the motion to the Senate. Penny had met with the Kungkas in April when she attended the afternoon tea at Parliament House in Adelaide.

The Senate:

(a) notes:

>(i) that 15 October, 2003 marks the 50th anniversary of the first atomic test conducted by the British Government in northern South Australia;

>(ii) that on this day "Totem 1", a 10 kilotonne atomic bomb, was detonated at Emu Junction, some 240 kms west of Coober Pedy;

>(iii) that the *Anangu* community received no forewarning of the test;

>(iv) that the 1984 Royal Commission report concluded that Totem 1 was detonated in wind conditions that would produce unacceptable levels of fallout, and that the decision to detonate failed to take into account the existence of people at Walatina and Welbourn Hill;

(b) expresses its concern for those Indigenous peoples whose lands and health over generations have been detrimentally affected by this and subsequent atomic tests conducted in northern South Australia;

(c) congratulates the Kupa Piti Kungka Tjuta - the Senior Aboriginal Women of Coober Pedy, for their ongoing efforts to highlight the experience of their peoples affected by these tests;

(d) condemns the Government for its failure to properly dispose of radioactive waste from atomic tests conducted in the Maralinga precinct; and

(e) expresses its continued opposition to the siting of a low-level radioactive waste repository in South Australia.

Opposite page: A commemorative postcard was released on October 15. Designed by Georgina Wright, in collaboration with the Kungkas. Printed and distributed by Avant Card.

Right: IMAGINE (Melbourne under the Maralinga bomb). This image appeared on the back of the postcard.

Left: Through an interpreter Karina Lester and Nina Brown share the Irati Wanti story with thousands of German protesters.

Wir stellen uns quer—We will always be in the way

November

Karina Lester and Nina Brown flew to Germany to witness community resistance surrounding the annual nuclear waste transport. A local Farmers Emergency Committee and the Lüchow-Dannenberg Citizens Initiative for Environmental Protection funded their visit.

Every year the German Government mobilise 15,000 riot police (at a cost of $21 million AUD) to escort the dangerous nuclear cargo by train in heavy containers called Castors. The waste is transported from a French reprocessing plant to a temporary storage facility near the small rural village of Gorleben. Despite the culture of fear created by occupying police forces, people from all walks of life participate in creative, non-violent direct action.

From tractor convoys and street kitchens to road damage and tree houses, the two week tour was a crash course in diverse tactics employed by residents to protect their future from nuclear waste. Many people were also keen to hear about Irati Wanti and the expansion of Australia's nuclear industry, and Karina and Nina spoke publicly on many occasions including; 5000 people at a protest, special public meetings, journalists and a senior school class.

Almost 200 tractors form a protective boundary around 5000 protesters in an open field near Splietau.

German police guard the radioactive transport through a small town.

Returning to Australia, Karina told the *Koori Mail* newspaper she travelled to Germany on behalf of her grandmother, Eileen Brown, and it was wonderful that A<u>n</u>angu were represented. "I was very proud that these people invited us and wanted to hear about our fight."

"We come home with much inspiration, grateful we're not planning blockades along the transport route but still have the chance to stop the dump once and for all," Nina said back in Coober Pedy.

Five hundred school students march to the police barracks.

Karina speaks to senior school students.

Karina at one of the many protest camps

Pilgrims from Australia and Japan join Eileen Brown as she prays for a safe journey.

Stepping out for peace

December

The Kungkas and Irati Wanti coordinators headed to Roxby Downs for the launch of the International Peace Pilgrimage. This eight month journey on foot would pass through Adelaide, Melbourne and Canberra before travelling on to Hiroshima, Japan.

Camped just kilometres from Olympic Dam uranium mine, the peace pilgrims prepared for the epic journey ahead. Around a fire they pledged, "Every step will be a prayer towards a nuclear-free future." The next day everyone convoyed to Olympic Dam, stretching banners and flags across the mine entry gates.

Atsuko Nogoua, pilgrimage organiser, and Jun Yasuda, a Japanese Buddhist monk, sing with the Kungkas at the mine gates.

Kungkas have a dip in the Roxby Downs swimming pool.

Left: Angelina Wonga ties a paper crane, a traditional symbol of peace, to the gates of Olympic Dam.

The Kungkas call on the pilgrims to follow them, leading the first steps of the walk destined for Hiroshima.

2004

"Get your ears out of your pockets"

February

Dr John Loy, CEO of Australia's nuclear regulator, the Australian Radiation Protection and Nuclear Safety Agency (ARPANSA), was empowered to grant or refuse the Federal Government's licence applications to prepare the site, construct, and operate the dump. Loy invited key stakeholders to present cases for and against the application in Adelaide at a two day forum.

The Adelaide forum followed a public consultation period, in which over 1000 submissions were made, with 99% in opposition to the dump.

On Day One the Kungkas joined Dr Roger Thomas from the Kokatha People's Committee, and representatives from the SA Government, Sutherland Shire Council, NSW Local Government Association, ACF, FoE and Democrats, to speak against the application.

The unfamiliar setting of the brightly lit stage, complete with podium, microphones, water carafes and a panel of three white men, didn't deter the Kungkas from *tjukaruru wangkanyi*—speaking directly. Ivy Stewart, Eileen Brown, Eileen Crombie, Emily Austin and Martha Edwards braved the formalities and began their presentation with *Inma*.

Eileen Crombie and Eileen Brown outside the Adelaide Convention Centre.

Karina Lester introduced the Kungkas and explained *Anangu* connection to the desert country. "It might look barren and very scarce out that way, but to us *Anangu* people it's very much part of our life." Karina asserted that just as ARPANSA have responsibilities for nuclear protection and radiation safety, the Kungkas also have responsibilities for their country and culture.

Each Kungka stood up and reiterated her deep opposition to the waste dump, all sharing their experiences of atomic testing in the 1950s and their strong concerns for the *manta*—land, the *kapi*—water and the *tjitji tjuta*—all the children.

Despite widespread cynicism that ARPANSA was simply conducting a rubber-stamping exercise, there was little doubt that Loy and fellow panelists Ian Lowe and George Jack were moved by the presentation.

Before the ARPANSA forum Eileen Crombie addresses media and supporters.

Letter to ARPANSA panelists on Day Two of the forum. Early morning the news hit of two isolated fatal truck crashes that had occurred overnight in the Blue Mountains, NSW.

"You all wake up now?"

February 26, 2004

Listen. You all wake up now? Everyone should know now. These trucks today shows them we know what we talking about. That's the one we was telling them over and over. It's dangerous. We've driven those roads, travelled through those mountains, we've seen it.

Now Government's saying it's safe, won't hurt nobody, but those trucks in the Blue *Apu*—Blue Mountains, show it's not safe. Shows the world it's true, not a lie, these words come from the heart. They got kids that way too in Blue *Apu*.

We know from our *kamis* and *tjamus*—grandmothers and grandfathers, they been telling us long time, like the Old Testament, old stories. Our *tjamus* and *kamis*, they didn't have books, they carried stories in their heads and told them from the heart. We got the *Inma*—ceremony, not in a book, in the *manta*—earth, in our heart.

The Kupa Piti Kungka Tjuta is sorry for the families of those men who passed away in that accident. Those truck accidents, two fellas lost their lives, family lost them now, missing those blokes. Through driving chemicals, that poison, for someone else to get rich. Stop it now, don't let them drive those trucks with poison, they've got family and kids too.

Yesterday, we told the ARPANSA mob that it's not safe, that truck might have accident. Today the Government gonna try and tell them it's safe. We know we're talking true way. You gotta listen. Get your ears out of your pockets and start using them.

Kupa Piti Kungka Tjuta

Letter to Dr Caroline Perkins, Department of Education, Science and Training. She presented the Federal Government's case for the licence application on Day Two of the ARPANSA forum. In her presentation she claimed Aboriginal groups had provided clearance for the dump.

February 26, 2004

"We never said yes"

To Caroline Perkins,

We never say okay. We don't want the dump. Everybody, all of us, we never say 'okay'. We got the kids' country. We're not going to give it away.

Stop telling lies that we agreed to the dump in the clearances. We always said 'NO', we're not going to give away. Why don't you come back and talk and listen to us face to face, sit down and talk to one another. But don't talk *ngunti*—in a lying way that we gave away about the dump.

We never said 'yes'. *Wiya*—No. We never said 'yes'.

Eileen Unkari Crombie, Emily Munyungka Austin, Eileen Kampakuta Brown, Martha Uganbari Edwards, Ivy Makinti Stewart.

Kupa Piti Kungka Tjuta

Before heading home the Kungkas caught up with greenies from Adelaide, Melbourne and Sydney for afternoon tea.

FEBRUARY

STATE INQUIRY CALLS FOR TRANSPORT BAN

A NSW Parliamentary Inquiry says plans for a radioactive waste dump "can't be justified" and a proposal to transport waste from the reactor should be abandoned.

The resulting report calls on the NSW State Government to ban radioactive waste transport through NSW. The Kungkas made a submission to the Inquiry in 2003.

The proposed waste transport route.

Talk'n Up Country

March

The Kungkas travelled south once again to participate in the Adelaide Festival of the Arts.

Emily Austin, Eileen Crombie, Eileen Brown, Myra Watson and Martha Edwards held two free public workshops sharing their cultural skills as part of the Talk'n Up Country program. At Tandanya Cultural Centre and on Universal Families Day they demonstrated methods used to make their beautiful bush crafts.

The workshops were interactive, and people had the opportunity to try their own hand at basket weaving and burning Ininti beans to make necklaces.

Eileen Brown teaches a young woman the art of basket weaving.

Eileen Crombie demonstrates the burning of intricate designs on a clapstick.

The Kungkas also catch a fantastic performance by WA's Bardi Dancers from island country north of Broome.

MARCH 18

PM DEFENDS DUMP SITE AS 'RIGHT SPOT'

"Everyone can find a reason why it shouldn't be somewhere, but the experts told us that the site identified was the most appropriate and therefore we have accepted their advice." John Howard, Prime Minister.

The Kungkas take the Germans to Coober Pedy's water source; an underground bore 22 kms out of town. Heike Lehmberg, Emily Austin, Eileen Brown, Hagan Voss, Magdalene Seffers-Michalski.

A troublesome trio

March

Three German farmers and anti-nuclear campaigners conducted a speaking tour along the proposed waste transport route. They concluded their tour in Coober Pedy where they met with the Kungkas. Magdalene Seffers-Michalski, Heike Lehmberg and Hagan Voss are from northern Germany, where annual high-level nuclear waste transports are met with vigorous community resistance.

The trio prompted a flurry of responses from the Federal Government. Peter McGauran, Science Minister, tried to discredit the Germans' speaking tour as "irrelevant", as there had been no nuclear waste accidents in Germany or Australia. In reply, Heike asked the Minister on national radio, "Do we really have to have a major accident before we can worry about this issue?"

While the international visitors stirred up Canberra, in Coober Pedy their stay was a great opportunity to return the hospitality enjoyed by Karina and Nina the previous year, providing a welcome burst of energy and solidarity.

APRIL 1

SCATHING REPORT FROM ARPANSA PANELIST

Dr Ian Lowe sat on the panel that the Kungkas presented to in February. He warns ARPANSA CEO, John Loy, that the Federal Government's dump proposal is "so clearly deficient" and that the evidence supplied raises "significant unanswered questions about the capacity of the Government to manage the project effectively to guarantee public accountability".

A special award ceremony

May

At Umoona Aboriginal Aged Care, a special ceremony was held in honour of Mrs Ivy Makinti Stewart.

On Australia Day 2004, Mrs Stewart received a Premier's Award for Outstanding Community Service but couldn't attend the official award ceremony in Adelaide. Aged Care residents, local Elders, and Mrs Stewart's family and friends gathered to recognise the achievements of the oldest founding member of the Kupa Piti Kungka Tjuta.

Mrs Stewart, who is in her 80s, received her medal and inscribed vase and she was proud to finally be recognised for all her hard work.

Eileen Brown and Ivy Stewart.

Karen Crombie, Eileen Crombie's granddaughter, hosts the ceremony.

Top: Eileen Brown, Ivy Stewart, Emsey Lennon. Bottom: Nelly O'Toole and Denise Lennon.

Leigh Cleghorn, Manager of Umoona Aged Care, proudly acknowledges Mrs Stewart.

MAY 11

BATTLE OF WILLS OVER LAND GRAB

The SA Government challenges the Federal Government's compulsory land acquisition of Site 40a in the Federal Court. While Peter McGauran, Federal Science Minister, says the SA Government is "delaying the inevitable and wasting taxpayer's money," John Hill, SA Environment Minister, says the SA Government will pursue "all available legal and political avenues to stop the dump".

Emily Austin and Eileen Brown, Bulgunnia Station.

JUNE 24

LAND GRAB FOUND ILLEGAL

The full bench of the Federal Court deems the compulsory land acquisition unlawful. The unanimous judgement in favour of SA finds the Federal Government had misused the urgency provisions to secure Site 40a for the dump.

"After 11 years and millions of dollars spent in selecting the very best site, the safest, the most secure, then it is nonsense to pretend there is any other site that's better [than 40a]... One way or another the dump will proceed at Woomera," vows Peter McGauran, Federal Science Minister.

July-August 2004

JULY 2004

CAN THE GOVERNMENT BACK DOWN NOW?

Despite Peter McGauran's confidence that the dump will proceed at Site 40a, his Government is running out of options. They can either challenge the court decision in the High Court which would cost millions and delay the process for months; or go through a normal compulsory land acquisition process without using the urgency provision, giving the SA Government ample time to pass the Public Parks Bill and turn Site 40a into a park.

The Federal Government are under fire with an October election looming and three Liberal seats in Adelaide at risk. Will John Howard jeopardise his re-election hopes for the waste dump?

JULY 7

HOWARD PUTS HIS FOOT IN IT...

The Prime Minister visits Adelaide to boost support for the Liberal Party during their re-election campaign. On Adelaide radio he admits the dump is "a no-win situation for the Government" and casts serious doubt on the future of the replacement nuclear reactor at Lucas Heights.

JULY 10

HOWARD FEARS POLITICAL FALLOUT...

After constantly being heckled by anti-nuclear protesters and unrelenting media attention on the dump issue, John Howard promises to "reconsider" the issue when the Federal Cabinet meets next week.

JULY 13

CABINET DEADLOCKED...

Federal Cabinet meets in Canberra and is split on the dump issue. The meeting includes long-term dump advocates: Nick Minchin (Finance Minister), Peter McGauran (Science Minister) and David Kemp (Environment Minister).

JULY 14

FEDERAL GOVERNMENT ABANDONS RADIOACTIVE WASTE DUMP PLANS FOR SOUTH AUSTRALIA

No dump for SA

On July 14, 2004 the Federal Government abandoned their plans for a radioactive waste dump in South Australia.

Due to massive community pressure they retreated from their multi-million dollar campaign, more than six years after identifying Billa Kalina as the 'ideal' region for the dump.

John Howard's media release stated that the Federal Government will start a new search outside SA for Commonwealth land to dump low and intermediate long-lived waste from the Lucas Heights nuclear reactor.

When the announcement hit news stands the Kungkas and the Irati Wanti coordinators were spread across the country:

Eileen Brown, Eileen Crombie, Angelina Wonga, Karen Crombie, and Clare Brown were out of contact at the Aboriginal Women's Law and Culture meeting near Ernabella in the state's far-north.

Nina Brown, in Melbourne for the Students for Sustainability conference, had presented an Irati Wanti workshop only the day before. She celebrated the Government's decision with many greenies who had been instrumental in the no-dump campaign. An Irati Wanti benefit already planned for that night at Irene Warehouse turned into a victory party.

Kungkas are 'out of range' in SA's far-north.

MKs join with over 250 greenies to dance the night away. Emily Johnston, Nina Brown, Camilla Pandolfini, Shannon Owen, Breony Carbines, Sam Sowerwine, Eve Vincent.

Happy greenies. Top: Eric Miller, Dimity Hawkins, James Courtney, Loretta O'Brien, Nina Brown. Bottom: Genevieve Rankin, Bruce Thompson, Dave Sweeney.

"We are winners because of what's in our hearts"

July 15

Emily Austin had heard the news over the radio while out bush near Kingoonya and she was overjoyed. When she returned to Coober Pedy the following day she celebrated the victory at a barbecue with her family and Georgina Wright.

Top: Sandra Taylor, Tiffany Brown, Pearl Austin, Anna Tsakaridis, Dimos Tsakaridis, Michael Austin, Vincent Warren, Isaac Warren, Norman Riessen, Robert Austin, Georgia Brown, Jonathan Riessen, Natalie Austin, Chuckisha Hayes, Samantha Hayes. Bottom: Emma Austin, Cherika Lang, Simon O'Toole, Douglas Waye, Emily Austin, Destiny Brown.

July 17

Homeward-bound from the Law and Culture Meeting, the Kungkas finally heard the news at Marla Bore Roadhouse. Cheering and clapping erupted from the Toyota Troupe Carrier. "Dump *wiya*—no more... KUNGKA TJUTA WINNERS." Cheerful singing continued for the next 250 kms down the Stuart Highway.

July 19

Eileen Brown, Emily Austin, Eileen Crombie, Angelina Scobie, Georgina Wright and Clare Brown travelled to Woomera. They met with Eileen Wingfield, her daughter Winnie and granddaughter Chantelle, Kokatha men Andrew and Robert Starkey, and 50 greenies on the annual Radioactive Exposure Tour to celebrate the victory and visit Arcoona Station.

Kungkas and greenies at the campsite near Woomera.

At Arcoona Station, no longer the planned dump site for Australia's radioactive waste. Angelina Scobie, Emily Austin, Eileen Wingfield, Eileen Brown, Eileen Crombie.

Greenies paint KUNGKA TJUTA WINNERZ banner in preparation for the Kungkas arrival.

Thumbs up for the Kungkas.

Bob Norton from nearby town of Andamooka, reflects on his efforts to stop the dump.

Eileen Crombie, Robert Starkey, Eileen Wingfield, Andrew Starkey, Eileen Brown, Emily Austin.

Kungkas and greenies convoy to Arcoona.

Eileen Brown and Eileen Crombie at Arcoona.

July 26

In Coober Pedy for a community cabinet meeting, SA Premier Mike Rann had morning tea at Umoona Aged Care and paid tribute to the Kungkas' strength and determination throughout their successful campaign.

Eileen Crombie, Eileen Brown, Premier Mike Rann, Emily Austin.

Right: Emily Austin presents Mike Rann with clapsticks.

Far right: Clare Brown, Marian Prickett, Eve Vincent and Mike Rann share a laugh over damper and a cup of tea.

"People said that you can't win against the Government. Just a few women."

August 6, 2004

People said that you can't win against the Government. Just a few women. We just kept talking and telling them to get their ears out of their pockets and listen. We never said we were going to give up. Government has big money to buy their way out but we never gave up. We told Howard you should look after us, not try and kill us. Straight out. We always talk straight out. In the end he didn't have the power, we did. He only had money, but money doesn't win.

Happy now. Kungka winners. We are winners because of what's in our hearts, not what's on paper. About the country, bush tucker, bush medicine and *Inma*. Big happiness that we won against the Government. Victorious. And the family and all the grandchildren are so happy because we fought the whole way. And we were going away all the time. Kids growing up, babies have been born since we started. And still we have family coming. All learning about our fight.

We started talking strong against the dump a long time ago, in 1998 with Sister Michele. We thought we would get the greenies to help us. Greenies care for the same thing. Fight for the same thing. Against the poison.

Since then we been everywhere talking about the poison. Canberra, Sydney, Lucas Heights, Melbourne, Adelaide, Silverton, Port Augusta, Roxby Downs, Lake Eyre. We did it the hard way. Always camping out in the cold. Travelling all over with no money. Just enough for cool drink along the way. We went through it. Survivors. Even had an accident where we hit a bullock one night on the way to Port Augusta. We even went to Lucas Heights reactor. It's a dangerous place, but we went in boldly to see where they were making the poison, the radiation. Seven women, seven sisters, we went in.

We lost our friends. Never mind we lost our loved ones. We never give up. Been through too much. Too much hard business and still keep going. Sorry Business all the time. Fought through every hard thing along the way. People trying to scare us from fighting, it was hard work, but we never stopped. When we were going to Sydney people say, "you Kungkas cranky, they might bomb you", but we kept going. People were telling us that the whitefellas were pushing us, but no, everything was coming from the heart, from us.

We showed that greenies and *A<u>n</u>angu* can work together. Greenies could come and live here in Coober Pedy and work together to stop the dump. Kungkas showed the greenies about the country and the culture. Our greenie girls are the best in Australia. We give them all the love from our hearts. Family you know. Working together, that's family. Big thank you to them especially. We can't write. They help us with the letters, the writing, the computers, helped tell the world.

Thank you very much for helping us over the years, for everything. Thank you to the Lord, all our family and friends, the Coober Pedy community, Umoona Aged Care, the South Australian Government and all our friends around Australia and overseas. You helped us and you helped the kids.

We are happy. We can have a break now. We want to have a rest and go on with other things now. Sit around the campfire and have a yarn. We don't have to talk about the dump anymore, and get up and go all the time. Now we can go out together and camp out and pick bush medicine and bush tucker. And take the grandchildren out.

We were crying for the little ones and the ones still coming. With all the help we won. Thank you all very much.

No radioactive waste dump in our *ngura*—in our country!

Kupa Piti Kungka Tjuta

Mrs Ivy Makinti Stewart

Mrs Eileen Kampakuta Brown

Mrs Angelina Wonga

Mrs Emily Munyungka Austin

Mrs Eileen Unkari Crombie

Mrs Tjunmutja Myra Watson

Image Credits

© All images copyright of the individual photographers.
c/o IW: copyright remains with the Irati Wanti campaign as photographers are unknown.

Front cover: c/o IW
Pg 1: *Simpson desert trees.* Clare Brown
Pg 3: *Tjitjis at Ten Mile.* Nina Brown
Pg 4: *Coober Pedy.* Clare Brown
Pg 5: Nina Brown
Pg 7: *Portraits (from top left).* Nina Brown; Nina Brown; Nina Brown; Lesley Johns; c/o IW; Clare Brown; Clare Brown; Emily Johnston; Peter McConchie; c/o IW; Peter McConchie; c/o Karina Lester
Pg 9: *Sunset at Mabel Creek.* Clare Brown
Pg 10: *(from top)* c/o IW; *A Radioactive Waste Repository for Australia: Methods for choosing the right site.* National Resource Information Centre
Pg 11: Clare Brown; c/o IW
Pg 13: Camilla Pandolfini
Pg 14: *(from top)* Clare Brown; Michele Madigan; Nina Brown
Pg 15: c/o IW
Pg 16: Nina Brown
Pg 17: c/o IW
Pg 18: c/o IW
Pg 19: *Emu tjina—footprints.* Clare Brown
Pg 20: Kelly Warren
Pg 21: *(from top) Simpson.* Nina Brown; *Breony.* Clare Brown; *Tracker flyer.* c/o Melbourne Kungkas; *In memory.* Breony Carbines; *St Andrews.* Breony Carbines; *Irene Warehouse.* Breony Carbines; *Picking bush medicine.* Nina Brown; *Playing cards.* Nina Brown
Pg 22: *Opal Festival.* Indira Narayan; *Radioactive Exposure Tour.* Bruce Thompson
Pg 23: c/o Humps not Dumps
Pg 24: c/o Humps Not Dumps; *Movie star at Ten Mile.* c/o IW
Pg 25&26: Fernando Gonzales
Pg 27: *Desert scrub.* Clare Brown
Pg 28: Clare Brown
Pg 29: Emily Johnston
Pg 30: *Portraits (from left)* Nina Brown; Clare Brown; Nina Brown; Nina Brown; George Wright; George Wright
Pg 31: *(from top left) Digging.* Nina Brown; *Happy drum.* Nat Walsey; *Clare and Georgia.* George Wright; *Eileen Wingfield and Lucy.* c/o IW; *EJ, Boom box and Coppa.* Francesca da Rimini; *Station Day.* Amira Pyliotis; *Clare computer.* Emily Johnston; *Digging for maku.* Clare Brown; *Sleeping.* Nina Brown; *Greenie Girls.* Peter Matthews; *Michele Warren action.* David Noonan; *Clare cooking.* George Wright
Pg 32: *(from left) Opal Festival.* Breony Carbines; c/o IW; *(from left) Kungkas in the capital.* c/o IW; Michele Madigan
Pg 33: *Lake Eyre sign.* Bruce Thompson; c/o IW
Pg 34: Marta Iniguez
Pg 35: *Race Around Oz.* Clare Brown; *Olympic Vision.* Lucy Brown
Pg 36: Lucy Brown
Pg 37: Lucy Brown
Pg 38: Michele Madigan
Pg 39: Michele Madigan; *Irati Wanti online.* Nina Brown
Pg 40: Clare Brown; *Back of Irati Wanti video:* c/o Irati Wanti Office
Pg 41: *Cadney grasses.* Clare Brown
Pg 42: c/o IW
Pg 43: c/o IW
Pg 44: Margaret Mackay
Pg 45: Clare Brown
Pg 46: c/o Alliance Against Uranium; *Yeperenye.* George Wright
Pg 47: *Waking up with Dreaming.* Peter McConchie; *Protest threads.* Clare Brown
Pg 48: c/o IW
Pg 49: *Breakaways.* Camilla Pandolfini
Pg 50: *(from top left)* Emily Johnston; Emily Johnston; Francesca Da Rimini; Francesca da Rimini
Pg 51: *Looking after culture.* Francesca da Rimini and Emily Johnston
Pg 52: Lucy Brown
Pg 53: Clare Brown
Pg 54: *Avant card.* c/o Irati Wanti Office; *National radioactive waste repository Draft EIS.* PPK Environment & Infrastructure
Pg 55: George Wright
Pg 56: Bruce Thompson

Pg 57: *Desert rocks.* Clare Brown
Pg 58: Benjamin Storrier
Pg 59: Benjamin Storrier
Pg 60: *(from top)* Emily Johnston; Clare Brown
Pg 61: *(from top)* Nina Brown; c/o Eileen Brown
Pg 62: *(from top left)* c/o IW; Nina Brown; Nina Brown; Michele Madigan
Pg 63: *(from top)* Goldman Environmental Prize; c/o Eileen Wingfield
Pg 64: *(from top)* Jodie Patterson, courtesy of Koori Mail; Michele Madigan
Pg 65: *(from top left)* Nina Brown; Michele Madigan; Michele Madigan; Michele Madigan; *Sutherland Shire.* Emily Johnston
Pg 66: Clare Brown
Pg 67: *Living Black.* Clare Brown; *National radioactive waste repository Draft EIS.* PPK Environment & Infrastructure
Pg 68: Francesca da Rimini
Pg 69: Sally Beaumont
Pg 70: Francesca da Rimini
Pg 71: *Arcoona sign.* Bruce Thompson
Pg 72: Clare Brown
Pg 73: Clare Brown
Pg 74: c/o Irati Wanti office
Pg 75: *Kulini Kulini poster.* c/o Irati Wanti office; *Japanese professor.* George Wright
Pg 76&77: *The making of Bush Camp.* assorted c/o Irati Wanti office
Pg 78: Clare Brown
Pg 79: Peter Matthews
Pg 80: *(from top)* Clare Brown; George Wright; Marta Iniguez
Pg 81: *(from top)* Peter Mathews; Peter Mathews; Marta Iniguez
Pg 82: *(from top)* Peter Mathews; Peter Mathews; George Wright; Marta Iniguez
Pg 83: *(from top left)* George Wright; George Wright; Marta Iniguez; Suzanne Woolford
Pg 84: *(from top)* c/o IW; George Wright; George Wright
Pg 85: *(from top left)* George Wright; George Wright; George Wright; Clare Brown
Pg 86: *(from left)* Marta Iniguez; Marta Iniguez; Clare Brown
Pg 87: Clare Brown
Pg 88: Clare Brown
Pg 89: *(from top left)* George Wright; George Wright; Clare Brown
Pg 90: Peter Matthews
Pg 91: Peter Matthews
Pg 92: Peter Matthews
Pg 93: *(from top)* c/o IW; Breony Carbines
Pg 94: c/o Irati Wanti office
Pg 95: *Design by George Wright.* c/o Irati Wanti office
Pg 96: c/o IW
Pg 97: *(from top left)* Nina Brown; Nina Brown; Magdalene Seffers-Michalski; Nina Brown
Pg 98: Nina Brown
Pg 99: *(from top left)* Nina Brown; Nina Brown; Emily Johnston; Nina Brown
Pg 100: Nina Brown
Pg 101: *Moon rise highway.* Clare Brown
Pg 102: Lesley Johns
Pg 103: Lesley Johns
Pg 104: c/o IW
Pg 105: Clare Brown
Pg 106: George Wright
Pg 107: Clare Brown
Pg 108: Nina Brown
Pg 109: *Water in the desert.* c/o IW
Pg 111: Nina Brown; Karrina Nolan; Nicky Forster
Pg 112: George Wright
Pg 113: *(from top)* Donna Green; c/o IW,
Pg 114: *(from top left)* c/o IW; George Wright
Pg 115: Mark Thompson
Pg 117: c/o Irati Wanti office
Pg 120: *Sun over Mabel Creek.* James Brown
Back cover: c/o IW